DISTINGUISHED MEN AND WOMEN OF SCIENCE, MEDICINE AND THE ARTS

THE INVENTIONS AND DISCOVERIES OF THE WORLD'S MOST FAMOUS SCIENTISTS

DISTINGUISHED MEN AND WOMEN OF SCIENCE, MEDICINE AND THE ARTS

Additional books in this series can be found on Nova's website under the Series tab.

Additional e-books in this series can be found on Nova's website under the eBooks tab.

DISTINGUISHED MEN AND WOMEN OF SCIENCE, MEDICINE AND THE ARTS

THE INVENTIONS AND DISCOVERIES OF THE WORLD'S MOST FAMOUS SCIENTISTS

MICHAEL F. SHAUGHNESSY

AND

MANUEL F. VARELA

nova
science publishers
New York

Library of Congress Cataloging-in-Publication Data

ISBN: 978-1-53613-282-3

Published by Nova Science Publishers, Inc. † New York

This book is dedicated to my loving wife, Ann Higgins Varela, and my two supportive children, Timothy and Edward.

—M.F. Varela

CONTENTS

PREFACE

Who were those famous men and women of science who gave us so much? What were they like? What motivated them and what led them to their amazing discoveries? This book takes us on an amazing journey through the hearts, minds and souls of so many scientists, inventors and scholars who have given civilization so much over the centuries.

Many of these scientists had an idea. They had some sort of insight or inspiration when they observed the world about them and decided to make it better. They had humankind in mind, and they set about their work to improve the human condition. They decided to invent. They decided to explore. They decided to investigate and reveal some of nature's deepest and darkest secrets to humankind.

Some of these scientists had to overcome gargantuan obstacles to reach the pinnacle of their craft and success. Some had to work endlessly to develop their insights. Some worked with visible elements, while others delved into realms that cannot be seen by the naked eye. Many of these scientists were recognized by their peers and traveled to Gamla Stan in Stockholm, Sweden to receive the highest of honors, the Nobel Prize in their field. Some of these scientists' names are forever etched in our memories; Gregor Mendel, Marie Curie, and others will always be noted and recognized for their foundational work, which has given humankind so much. Other scientists and their discoveries have laid the foundation for future work in the realms of medicine and other domains.

This book takes us on a journey so that we can learn more about the lives and passions of the most famous scientists from the past and present. Their foibles, their strengths and their passions will be revealed and discovered in this text, just as their discoveries have opened the doors for their followers and students that they have mentored over the years.

We can learn much from the lives of these famous scientists; not just about their studies, but the way in which they conducted their lives and from the contributions that they gave to others, their countries and to society at large.

We are about to undergo an adventure. An adventure into the minds of the world's most famous scientists and inventors; we will be guided by Dr. Manuel F. Varela who will share his own knowledge and past encounters with some contemporary scientists, as well as some stories that make their famous people more human and caring, and more than just a formula.

We must acknowledge several people in this endeavor, including Ann Varela, Dr. Ken Cradock and other anonymous reviewers. Furthermore, we have to thank Jimmy Kilpatrick of EducationViews as many of these interviews appeared first in that online journal. And now, the journey begins: A journey not just of interest, but of hope. Hope for a better future for all humankind via the discoveries and the inventions of the world's greatest scientists.

Michael F. Shaughnessy
Manuel F. Varela

Chapter 1

FRANCESCO REDI

1) Professor Varela, I would like to ask you some questions about Francesco Redi and his contributions to science. Can you provide some background as to his birth (when and where) and his early endeavors and education and experience?

Thank you for inquiring about the amazing scientist Dr. Francesco Redi. He was born in the town of Arezzo, Italy, on February 18, 1626. In his early years, he was educated in Florence at a Jesuit-run school, which taught mainly theology. When Redi turned about 15 years of age, he moved to the University of Pisa, where he ultimately earned his doctorate degrees in medicine and philosophy in 1647. Dr. Redi subsequently became the personal physician to the Grand Duke of Tuscany, Ferdinand II, and later to his successor, Cosimo III.

2) Now, I may be wrong, but Redi was apparently the first to test Aristotle's "spontaneous generation" ideas - using flies - How did this all come about and why is it important in the realm of science?

You are certainly correct about Dr. Redi being the first scientist to experimentally challenge the idea of spontaneous generation. In fact, Dr. Redi has the distinction of apparently publishing one of the very first, if not the first, experimentally-based paper dealing with the biological sciences, complete with proper controls.

The concept of spontaneous generation originated with Aristotle back in 343 B.C., when he postulated that living beings could simply generate themselves into existence spontaneously out of non-living materials, a notion that is to this day referred to as abiogenesis (i.e., life arises out of non-life). It is generally believed that Aristotle himself actually invoked, as an example, the flies as spontaneously generating themselves straight out of meat that's left to sit out, open to the air.

Although at first glance one might erroneously envisage how this might somehow be true, being that if one does leave meat exposed to air, one might certainly see flies "emerging spontaneously" from the meat. Incidentally, according to the theory, air was believed to be a "vital force" that was necessary for spontaneous generation to manifest itself. That is, air was critical to the process.

The facts need to be taken into account, however, that flies will actually land on the exposed meat, lay their eggs and hatch, and be mistakenly interpreted by observers to mean that spontaneous generation has occurred in which flies arise directly from the non-living meat. This incorrect spontaneous generation concept was nonetheless widely believed for many centuries to be completely true, believed even well into the 19th century. Spontaneous generation is certainly not true, and Dr. Redi was the first individual in history to provide experimental evidence against it.

3) What other contributions was he known for and why is this one still studied?

Dr. Redi is known also in scientific circles for his work as an early pioneer in the field of parasitology. He studied a particular type of parasitic worm known as the helminth, a kind of tapeworm. Worm parasites, such as the helminthes, infect the guts of animals, including those of humans and thrive at the expense of their unfortunate animal hosts. Parasites are unable to live totally on their own, and they thus require a living host that has all the biological workings necessary for life and which a parasite lacks. Thus, parasites which infect a host benefit in terms of growth while their hosts, in turn, suffer from the loss of nutrients, etc. Lastly, one thing I neglected to mention is that in some of the literature dealing with scientific history, Dr. Redi is considered to be the father of parasitology.

4) As I recall "dropsfilia" (fruit flies) were central to his methods- why is this important for budding scientists to know about and why is it so critical?

Interestingly, Dr. Redi read the ancient works of Aristotle and was aware of his spontaneous generation theory, along with the putative involvement of the flies, meat, and air. Redi also read the publications of both Hooke and Leeuwenhoek. So, Redi was especially aware of the observations that were made about microbes and other specimens by these earlier investigators using their scopes. Furthermore, Redi knew about the writings of William Harvey, who essentially refuted the theory. In particular, however, Redi was intrigued by Leeuwenhoek's published descriptions of flies at the microscopic level; there was no doubt that their body structures were complex, leading Leeuwenhoek to reason, thus, that flies reproduced sexually by laying eggs, which would hatch and then release progeny flies. This reproduction theory, however, flew in the face of the firmly established and ongoing spontaneous generation theory.

For instance, if flies spontaneously generated themselves, then there was then no apparent need for sexual reproduction and, consequently, no need for their complicated body structures seen in the flies. In summary, the questions were as follows. Did flies reproduce sexually, or did they simply spontaneously generate? These were the main issues concerning Dr. Redi and his historically famous work.

In 1668, Dr. Redi described in his newly published paper how he put to the test the idea of spontaneous generation. He had set up several flasks, all with meat placed into each of them. In one flask the meat was sealed airtight, preventing access to flies and the outside air, with its "vital force." Another flask was covered with a gauze mesh that kept flies out but still allowed air exposure to occur upon the meat within the flask. The last flask was left totally uncovered and open directly to both air and flies. In the first two flasks, both open to air, no flies arose from the meat. In contrast, in the open flask which allowed air and flies open access, Redi described that he observed flies landing on the meat, likely laying their eggs, and the subsequent hatching of the fly eggs. Redi further noted that the fly larvae developed into young flies on the uncovered meat, and he further observed the flies simply flying away. Because no flies spontaneously arose out of the meat within either of the sealed or the gauze-covered flasks, Redi concluded, correctly, that spontaneous generation did not occur, at least in flies. This famous experiment is still introduced in virtually all modern textbooks of Microbiology. It is also used as a good example of how to put the modern scientific method of inquiry to work when testing new ideas.

5) What are some interesting facts or details about this person and his life and contributions?

It may interest you and our readers to know that Dr. Redi had other intellectual interests. Apparently, he was recognized as a talented poet. He also pursued independent studies in the areas of linguistics, language dialects, and literature.

6) What have I neglected to ask about this interesting scientist?

I think it is fascinating to consider how Redi's famous fly experiment was received in his time and thereafter. On the one hand, an interesting outcome of the famous Redi fly experiment is that few people at the time were convinced of the fallibility of the spontaneous generation theory. Many people still believed in it. You'll recall that the idea had been resolutely established and firmly ingrained since ancient times, with Aristotle. So it was difficult to just let the idea go. In fact, Redi himself still thought that microbes generated spontaneously, even if his flies didn't.

An additional confounding factor was the work of another investigator, John Needham, who tested the spontaneous generation theory in microbes. Unfortunately, Needham's work still showed microbes apparently thriving in mutton gravy broth even after having been heated with a direct flame. The unfortunate happenstance is that Needham had had contaminating heat-resistant microbes, providing inaccurate evidence in support of spontaneous generation.

On the other hand, in 1767 Lorenzo Spallanzani, a well-respected clergyman and scientist in his own right, proposed that the tiny microbes landed on the mutton gravy broth and used it to grow, thereby refuting the spontaneous generation theory. The microbes were simply floating around in the air. The controversy continued, setting the stage for the spontaneous generation theory to remain intact well into the 19th century, until the definitive experiments of the great Prof. Louis Pasteur.

Chapter 2

ANTON VAN LEEUWENHOEK

1) Anton van Leeuwenhoek is one of our greatest scientists. Can you tell us a little bit about where he was born and when, and his education and early experiences and perhaps what led him into science?

Anton (or Antony) Phillip van Leeuwenhoek, a Dutchman, was born in the town of Delft, The Netherlands, on October 24, 1632. Although he did not receive a formal higher education in science as many of his contemporaries did, the 16-year old Leeuwenhoek started out his education as an apprentice in order to be a draper of linen. He later became a full-fledged draper businessman. Endeavoring to be the good businessman by providing good-quality linen, Leeuwenhoek personally built tiny handheld elegant-looking metal microscopes with glass lenses in order to magnify and thus closely examine his linen, to make sure they were of the best quality possible. His lenses could enlarge an image about 200 times the normal image.

Leeuwenhoek actually ground the glass lenses himself and did a remarkable job of it. He assembled his finely ground lenses into small brass and silver apparatuses with adjustment knobs and a place to hold the linen specimens for viewing. He could view his linens with a clarity and such an enhancement in relative size that modern microscopists, and even microbiologists, marvel at to this day. It is, however, a common misconception that Leeuwenhoek invented the microscope. He did not. He merely built his own microscopes. While in his free time, although it is difficult to imagine how Leeuwenhoek had any spare time as he was also, for instance, a city councilman, a haberdasher, a court surveyor, and a chamberlain of the laws for the city, Leeuwenhoek also examined other things that he found in his surroundings using his microscopes.

His entry into the field of natural science was rather circuitous because Leeuwenhoek himself did not seek the regular pathway to science as his contemporaries had. Instead, it was through the intervention of a colleague, Regnier de Graaf, a member of the Royal

Society, who indiscreetly informed the Society's secretary, Henry Oldenburg, about Leeuwenhoek's amazing findings with his homemade scopes. Encouraged by de Graaf and Oldenburg to write-up his microscopic findings to the society, Leeuwenhoek, considered then a non-scientist, finally wrote a letter to the Royal Society where this historical report about the microbes was read in their hall to the distinguished members of the society. Thus, in 1637 Leeuwenhoek began what was ultimately to become a lifelong correspondence with the regular scientists of the Royal Society.

2) What was he most famous for or what was his greatest discovery? (I know many scientists discover more than one thing?)

Leeuwenhoek is best known for having been the first human being ever to observe with his own eyes and consequently record the presence of the tiny microbial beings, which he called tiny "animalcules." When viewed through the lenses of Leeuwenhoek's scopes, he thought they were little animals. He was initially shocked and later astonished by these little beings.

He saw them moving about in his scope. Humankind had never known that these tiny creatures existed until Leeuwenhoek had first seen them and wrote about it. Hence, Leeuwenhoek is widely considered by historians and scientists alike to be the discoverer of the microbes and thus, is famous for being the founding father of the field of microbiology within the biological sciences. This first microbial observation by Leeuwenhoek is his greatest discovery, and, in fact, it is universally considered to be one of the greatest scientific discoveries in all of recorded history.

3) Is there any interesting story about his life, or his discoveries or his interests?

There are several remarkable aspects to Leeuwenhoek's story. First, I think it is interesting to note that Leeuwenhoek was an intensely curious person, like many outstanding scientists are, 1 believe. It may interest you and your readers to know that Leeuwenhoek's study of the world around him led him to find microbes in many seemingly odd places, like rainwater, soil, teeth plaque, or feces, for instance. But it is noteworthy that Leeuwenhoek also examined items like hair, sperm, blood, skin, etc. He reported these findings to the Royal Society, as well.

Second, Leeuwenhoek was extremely secretive about how to construct his handmade microscopes. He never shared with anyone his instructions on how to grind the lenses or to assemble his apparatuses. It is thought that Leeuwenhoek constructed over 500 of his scopes, and several have survived to this day. It is thought that because of this secretive nature, scientists in later generations were unable to build microscopes for themselves and extend the field of microbiology. Thus, the interest in microbes fell by the wayside for many years and possibly set back the forward movement of science in this area.

Third, there is an interesting story about how Leeuwenhoek would host parties and feature his scopes in which his guests were invited to view their own teeth plaque, known in Leeuwenhoek's day as the "scruf of the teeth." His guests were often astonished and delighted, but more often they were quite shocked, and some of his guests actually fainted from the revelation of these tiny living creatures being present about in odd places.

Lastly, I think it is quite remarkable that Leeuwenhoek's handwritten illustrations to the Royal Society detailing his microbes are, in fact, quite accurate, especially his drawings of the bacteria published in 1684. Bacteria are often illustrated today quite similarly.

4) What have I neglected to ask?

I think it is fascinating to consider what has resulted from Leeuwenhoek's microbial discovery. For example, because of our awareness of the microbes, we have since learned that some of them cause infectious disease, ranging from mild to serious to deadly, while other microbes are responsible for good, such as in the preparation of food (e.g., bread, sauerkraut, pickles, ice cream, etc.) and drink (e.g., beer and wine). Because of our knowledge and study of microbes, other advances became possible, such as vaccines, genetic engineering, medicines, biotechnology, genome sequencing, sewage treatment, and advancements in agriculture, just to name a few. Because of microbes, it is possible to produce large-scale amounts of vitamins, plastics, fuel (like oil, diesel, and gasoline), and important industrial chemicals. It is also possible to use microbes to degrade contaminating materials in environmental spills, in a process called bioremediation. Lastly, it has since been discovered that microbes are likely responsible for life on Earth starting and continuing.

Without the microbes, our lives and the lives of all living beings on Earth could not have been possible in the first place and could not be possible today.

Chapter 3

ROBERT HOOKE

1) Robert Hooke is one of our foundational scientists. Can you provide some broad, global autobiographical information about him (where and when he was born, went to school, and the like?)

Thank you very much for your interest in amazing scientists and their stories. It is a pleasure to work with you on this project. According to a *Small Pocket Diary*, which was an autobiographical booklet popular in his day, Robert Hooke wrote that he was born on July 18, 1635 in Freshwater which is a peninsula located in the western end of the Isle of Wight, England.

After the death of his father, John Hooke, the 13-year old Robert used his inheritance funds to buy his way out of an art apprenticeship in order to enroll in Westminster School. Hooke later attended Christ Church College at Oxford, where became a laboratory assistant to a now famous chemist, Robert Boyle. Hooke subsequently became a professor of geometry at Gresham College where he remained for the rest of his life.

2) Now, what is he MOST famous for, what would you say are his outstanding contributions to the field of science?

A discovery for which Hooke is most famous for is his coinage of the term "cell" which were tiny enclosed structures that he observed under the microscope when viewing cork slices, and which biologists 175 years later invoked to describe the "cell theory," that is, that living beings were composed of these tiny cells. Along this line, in 1665 Hooke had published his famous book, called *Micrographia*, in which he included elegant illustrations of living organisms, primarily plant and entomological in nature. The study of modern cell biology owes its origin to Hooke's cellular notion. Today, the field of cell biology is a central discipline within the biological sciences. Additionally, the workings of the living cell are critical for our understanding of many aspects in the biomedical sciences, such as in cancer biology, immunology, physiology, and molecular biology.

3) As with most scientists, there is some interesting story about his life, and his work. What has been very distinctive about this "father of cell biology"?

One interesting story that is often related to readers in biographical works of Hooke is his famous, or perhaps infamous, feuds with others, such as with the great Isaac Newton, over priority of ideas. The main premise of the feud between Hooke and Newton is grounded basically in who deserved credit for the mathematical discovery of gravitational mechanics associated with the movement of the planets. The feud between these two giants was persistent and became so contentious such that, at one point, an irate Newton, as head of the then Royal Society of London, arranged for the removal of the previous head's documentation and, sadly, an oil portrait of Hooke from the Society's building. Consequently, no image apparently exists of what Hooke actually looked like.

Another controversy arose with respect to who deserved the credit for the discovery of the microbes. Was it Antony van Leeuwenhoek or Robert Hooke? When one sorts through all of the arguments one way or another, historians have come to a tentative consensus that it was probably Hooke who may have first observed the bacteria (although he didn't publish this finding first) and Leeuwenhoek who first discovered the protozoa. If you hold that the discoverer who gets the credit is the one who published first, then the credit goes to van Leeuwenhoek for having properly published his observations of both types of microbes.

4) What have I neglected to ask about this famous scientist?

It is interesting to note that Hooke was a renaissance man in the sense that he is known also for having made contributions to many other distinct areas of expertise. For instance, to engineers, Hooke is known for his formulation of the so-called spring theory of how coiled springs contain their elasticity. To surveyors, Hooke was instrumental in helping to design the rebuilding of London after its devastating fire of 1666. To architects, Hooke had been known for his design of a new Royal College of Physicians with its then modern anatomy theatre, in 1670, after the original school had been destroyed during the Great Fire of London. Hooke later designed a large hospital, called Bethlehem or Bedlam, a grand townhouse, and many private homes, few structures of which have apparently survived.

One prominent surviving structure is the Monument to the Great Fire, designed in collaboration with architect Christopher Wren. To inventors, Hooke was known to have developed his own designs for improvements to a spring-balanced pocket watch, a barometer and a microscope. To physicists, Hooke is known for his theories pertaining to optics, such as his light and wave theories, that arose as a result of his work with the microscope and the telescope.

Chapter 4

Antoine Lavoisier

1) The name Antoine Lavoisier is linked with a great many ideas, theories and important aspects as we shall see. What do we know about his childhood?

The investigator Antoine Laurent de Lavoisier is known amongst the present chemists as the principle founder of the modern field of science called chemistry. Lavoisier was born on the 26th of August in the year 1743 in Paris, France, to parents Jean Antoine (his father) and Jeanne Emilie Punctis Lavoisier (his mother). Antoine's father was a prominent attorney, and his mother was herself the sole daughter of a wealthy attorney. The immediate Lavoisier family was known to be wealthy and bourgeois members of France's powerful nobility and aristocracy.

His mother Emilie died when Antoine was only 5 years of age, leaving a rather sizeable sum of money as an inheritance. He was then sent to live with his doting aunt, Marie Marguerite Constance Punctis (Aunt Constance), and his grandmother Punctis from his mother's side of the family, during this time onward. Some sources say that his widower father and kids lived together in the same household as his aunt and grandmother. Young Antoine had a younger sister, Marie Marguerite Emilie, who was born during the year 1745 and later died at the age of 15 years, bringing the close-knit family even closer together.

Antoine's devoted Aunt Constance, who never married, passed away later in 1781. She left the grand bulk of her own rather enormous wealth to Antoine, whom she had essentially raised ever since his mother had died. Additionally, his grandmother Punctis died in 1786 and left Lavoisier another hefty sum for his inheritance.

2) Apparently, when he went to College Mazarin, the procedure in terms of education then was some Professor came in and lectured- and then some grad or doctoral student would conduct some experiment proving what the grand old Professor said- but what happened with Lavoisier?

In 1754, when the child Lavoisier was approximately 11 years old, his father had enrolled him (Antoine) in the prestigious College de Quatre Nations, founded by Cardinal Mazarin, also referred to as the Mazarin College, where he was provided an education in the liberal arts and sciences, focusing primarily on the languages because, at the time, the young Lavoisier had initially wanted to become a writer. A short while later, when Lavoisier had changed his primary interests to those of the sciences, he nonetheless went into the study of law, receiving his undergraduate degree and law license in 1764, at the age of 21 years.

While it has been alleged that Lavoisier was greatly influenced by several of his professors while in college, such as by Prof. Abbe la Caille (astronomy & math), who had discovered the arc of the meridian around the Cape of Good Hope, by Prof. Guillaume-François Rouelle (chemistry), by Prof. Bernard de Jussieu (botany), and especially by Prof. Etienne Condillac (chemistry), he (Lavoisier) was, at the same time, not easily persuaded by the professors' proclamations during their lectures. During this period at Mazarin, the professors' lectures were often accompanied by an actual experimental demonstration of the proclaimed concepts, often conducted by a graduate student or a doctoral candidate, also known as a 'demonstrator' of the lectured concepts. Occasionally, a demonstrator's public 'demonstration' was in stark contradiction to the stated 'facts' put forth by the great Professor, and Lavoisier was known to have wisely and acutely been aware of the inconsistent nature of these lectured concepts versus the laboratory demonstrations.

Apparently, many of his fellow students merely took their notes obediently, seemingly unaware of these disparities. Lavoisier saw the need for careful experimental analysis and constant reevaluation of the commonly accepted notions held by his generation, especially when dealing with the concepts of alchemy. Lavoisier often had hope for successive generations in the acceptance the new ideas emerging from his own work.

3) Apparently, he was first involved in literature, then agriculture- what were his contributions to that field?

Lavoisier was made aware of a serious agricultural problem occurring in France and elsewhere in Europe and the New World pertaining to a disease known as the ergot poisoning, or ergotism. Although unknown to Lavoisier and his contemporaries at the time, certain grain crops, especially rye, may be contaminated with a fungus, now known by its scientific name *Claviceps purpurea*, producing a rye blight. The fungus produces a toxin, a derivative of ergoline-based molecules, that conveys upon their unsuspecting victims rather startling symptoms, like extreme pain in the extremities—patients reported that their hands and feet felt like they were on fire—plus, hallucinations and physical uncontrollable convulsions, among other non-specific signs and symptoms.

The ergot poisoning was known historically in the Middle Ages as the so-called "Holy Fire" and later as the "St. Anthony's Fire" because, as it has been recorded, if the patients made a pilgrimage to the church of St. Anthony's, they would find relief that was considered nothing short of miraculous. Historians and scientists have speculated that the monks at St. Anthony's fed their pilgrims rye bread that was not contaminated with the fungus.

It's been further postulated that the ergot poisoning also played a role in conferring behavior reminiscent of having been bewitched in those individuals who had been accused of practicing witchcraft in Salem, Massachusetts, during the 1660s.

Lavoisier became a member and secretary of a Royal Commission dealing largely with agricultural issues in France. Lavoisier was known to have spent a great deal of his time (about 10 years) and money to study and develop new farming practices devoted to not only improving crop yield but also to minimizing crop losses due to the ergot fungus and the rye blight. Lavoisier noted an association between unusually wet and rainy seasons with the onset of the rye blight and the ergotism.

Lavoisier was involved in the filing of a report addressing the rye blight issue. In the Royal Commission's report, Lavoisier provided an overview of certain farming practices that might possibly be implemented in order to solve the rye blight problem. Lavoisier suggested that other crops be farmed, such as those less susceptible to the ergot. He noted, however, that because the farmers were terribly poverty-stricken as a result of heavy taxing and unusually high rent rates, the potentially useful farming practices would necessarily be unrealistic. Thus, reform was essentially impossible.

4) Okay—key word—phlogiston — why is this important and why linked to Lavoisier?

During Lavoisier's time, it was widely believed that fire released a substance called phlogiston, contained within combustible materials. Lavoisier conducted experiments aimed at measuring the substances released during combustion. First, Lavoisier set the metal tin on fire and measured the weight of the resulting ash produced; he noticed the tin ash weighed more than the unburned tin! This was an unexpected result because it meant that during the fire the tin was picking up something, rather than losing something, like the putative phlogiston, perhaps. Next, Lavoisier repeated his combustion experiments but with the element phosphorous and found essentially the same sort of results, namely, that the burned phosphorus was heavier than it had been prior to the setting of it on fire.

Then, Lavoisier tried his combustion experiments with mercury. Again, Lavoisier found that the burned substances were heavier after being burned than they had been prior to being set on fire, suggesting that the combusted mercury material was actually picking up substances, rather than releasing them, like the so-called phlogiston theory had so eloquently predicted.

Lavoisier's work went totally against the widely held phlogiston concept—in a word, he became an ardent 'antiphlogist.'

5) Air consists, (and correct me if I am wrong) of oxygen and nitrogen. Why is this important in the big scheme of things- and how does it relate to Lavoisier?

You are certainly correct about the essentially non-polluted elemental composition of the Earth's atmospheric nature. The elements oxygen and nitrogen played a large part in the studies of Lavoisier.

Following up on his phlogiston experiments, Lavoisier found that combusted materials that he had set on fire were picking up oxygen and very likely nitrogen after the burning processes.

He measured the volume of the so-called 'dephlogisticated' air around the combusted materials using a large bell-shaped jar. The air volume had been reduced during the combustion process, suggesting that something from the air was being picked up by the burned materials. At first, Lavoisier called one of the new substances 'azote' which we now know to be the element nitrogen.

At first, critics of Lavoisier's antiphlogiston work maintained, albeit incorrectly, that the reduced air volume was indicative of and represented by the phlogiston. Favorable to Lavoisier's antiphlogiston work was the then recent discovery and purification of the element called oxygen by Joseph Priestley and Carl Wilhelm Scheele in 1774.

Using tin and mercury for his experiments, Lavoisier burned these metals as before, except that he then used a tightly sealed glass container while permitting the burned elements to become oxidized. Upon measuring the weight of the sealed vessel, he found no difference in amounts before and after their burning. However, when Lavoisier opened the sealed container, he made the astute observation of air rushing into the newly opened combustion vessels. Lavoisier reasoned that a vacuum had been generated in the sealed vessel, probably as a result of the burned material picking up a substance from the air, leaving behind the vacuum. Lavoisier deduced that, instead of phlogiston becoming liberated by combustion, a substance was taken up. This substance he later found to be present in acids. Therefore, in 1778, Lavoisier called this absorbed substance *oxygéne*, meaning "generator of an acid," or what we now know to be the element oxygen.

This discovery was to have a profound effect on the history and the progress within the field of chemistry. His work meant that during combustion, oxygen was taken up into the equations involving the chemistry of fire. The work essentially changed the entire direction of the investigations pertaining to the chemistry discipline.

6) Heat, and combustion - why are these two words key to Lavoisier's thinking about chemistry?

In Lavoisier's experiments, he had carefully controlled the amount of heat needed to initiate the combustion process during the burning of his experimental materials. For example, in his mercury experiments, Lavoisier heated mercury contained within a large bell-shaped jar apparatus, turning the mercury into a reddish colored ash heap. The reddish ash heap weighed more than it had before while the air had less volume. He had found that the leftover air was nitrogen.

Next, he repeated his mercury-heating experiment, except that this time he used the red ash heap as a starting material and raised the amount of heat to even higher levels than he had previously. This time, however, he had regenerated the mercury and at the similar weight that it had been before, demonstrating the reversibility of the process. This reversibility should have been impossible, at least according to the phlogiston theory, which had quite clearly stated that the chemical reactions cannot be reversed.

These experiments conducted by Lavoisier essentially had the effect of forever discrediting the phlogiston hypothesis, thus paving the way for progress to occur in the burgeoning field of chemistry. This work was incredibly important when, for instance, Pasteur later proclaimed that living beings could also conduct chemistry during the respiration process.

During the process of wine fermentation, Lavoisier found that sugar was converted to bicarbonate gas and to ethanol, a chemical compound he called the 'spirit of wine.' Lavoisier hypothesized that fermentation and the process of putrefaction occurred by similar modes. He showed the involvement of carbon dioxide in the course of respiration. He also clearly delineated the nature of the so-called 'fixed air,' which had been studied by Joseph Black, as carbon dioxide. These studies were to be critical to the development of both organic chemistry and later of biochemistry.

7) Now, I have to tell you- for years of my life in school- I stared at that Periodic Table of the Elements in various classes. Was Lavoisier somewhat initially involved in that big Table??

As you no doubt know and were quite likely taught as you stared at The Table, the periodic table of the elements was meant to include only those substances which cannot themselves be broken down any further. Although, in modern times, we now know that these elements representing the atoms can indeed be further broken down into sub-atomic particles, like quarks and muons, etc.

Lavoisier had several important contributions regarding the formulation of periodic table of the elements.

First, in studying water in 1783, Lavoisier found that it was not elemental in his strictest sense of the word, as he found that water consisted of oxygen, plus some other substance, which he had named as hydrogen. Prior to this discovery, Henry Cavendish had called this hydrogen gas an 'inflammable air.' Lavoisier had burned Cavendish's inflammable air in the presence of oxygen and found water consisted of hydrogen and oxygen.

Regarding other elements, Lavoisier has been credited with astutely predicting, in 1878, the presence of silicon as an element. Additionally, he is credited with having discovered the element sulfur. He was known to have burned sulfur and to have combined it with oxygen. Further, it has been reported that he coined the name carbon to this important element. Interestingly, the elements carbon, hydrogen and oxygen are all present in carbohydrates.

He was reported to have written the very first list of the elements. In 1789, after his death, his grand book was published, due largely through the efforts of his widow, Marie-Anne Lavoisier. The book was titled '*Elementary Treatise of Chemistry*' in which Lavoisier simply began making a listing of the various pure substances, the elements. Within the book, Lavoisier called this list a 'Table of Simple Substances.' Some historians of chemistry consider this book to constitute one of the first, if not the first, of the modern chemistry textbooks.

8) Elementary chemical nomenclature is attributed to him and a few colleagues- what is the story here?

Lavoisier had apparently made the suggestion that while the so-called simple non-decomposable substances be referred to as elements, the combination of any given elements, however, should be henceforth called compounds. Thus, the term chemical compounds owes its origin to Lavoisier and is commonly used to this day. Furthermore, he said that if a substance was known to undergo a chemical reaction with oxygen, the substance should be referred to as having been 'oxidized,' calling the new chemical variation an 'oxide.' Thus Lavoisier is recognized as having an important and long-lasting influence upon the nomenclature nature within the study of the chemical substances.

9) Conservation of matter in a specific chemical equation- why was this important?

Prior to Lavoisier's involvement with the phlogiston hypothesis, the principle of the conservation of matter had been put forth by Mikhail Lomonosov in 1756. Lavoisier's experimental studies with the anti-phlogiston work was completely in line with the law of matter conservation in that the overall weights of both starting and ending materials remained largely unchanged.

The so-called law of conservation of matter is vitally important not only for chemicals, as noted in Lavoisier's studies, but also critical when considering energy. The mass conservation principle is widely applicable to the fields of chemistry, physics, engineering, mechanics, etc. Lastly, the work of Lavoisier in this area began transition from the end of alchemy as we know it to the start of modern chemistry, as it exists today.

10) Sadly, for whatever reason, like so many in France at the time, he faced the guillotine. What happened?

Unknown to him, the path to the guillotine started early on for Lavoisier.

First, as a result of the inheritances from his devoted Aunt and his Grandmother Punctis, Antoine became an independently wealthy individual, leaving him plenty of free time to pursue intellectual interests when he became an adult. It also provided needed funding to perform many of his important experiments. Unfortunately, however, it also provided a secure means for other new and rather dubious investments, one being an investment in a notorious tax-collecting firm. It is reported, however, that Lavoisier had indeed tried to institute tax true reform measures that were meant to actually help the poor, but to no avail.

During the period of the French Revolution, all of the supremely hated tax collectors and anyone even remotely associated with tax collecting, such as Lavoisier, and his father-in-law, were themselves collected (arrested) by the revolutionaries, given a highly questionable and obviously unfair, but speedy trial, all based on trumped up charges, and were summarily condemned to death. It is also told that part of his demise centered on his refusal to allow admittance of Jean-Paul Marat to the French Academy of Sciences, a slight that Marat never forgot and was later to exploit in an effort to destroy Lavoisier.

Thus, his minor business association, in addition to being a bona fide member of the aristocracy, of course, were also to constitute another important motive for the prosecution of Lavoisier. His many scientific contributions were summarily rejected and deemed insufficient by his persecutors for any sort of hope of redemption. Sadly, on the day of the execution of Lavoisier by guillotine on the 8th of May in the year 1794, he was apparently first made to stand and watch the beheading of his beloved father-in-law, just before undergoing the execution himself.

11) His wife carried on his work after his death. Any ideas about her contributions?

In 1771, when Lavoisier was already 28 years old, he married the then 13 year old Marie-Anne Pierrette Paulze, the daughter of one of his senior tax-collecting business associates. His new wife, Madame Lavoisier, was reported to have had an educational background in the sciences and was well studied in the arts and languages.

The story is told that the marriage of Lavoisier to the very young Marie-Anne was actually a favor to her father, a senior business partner of the tax collecting firm, *La Ferme Generale*, of which Lavoisier was an investor. Apparently, the 40-year old Count d'Amerval had proposed to young Marie-Anne and made it clear that if she refused, her father would be fired from the tax-collecting firm. Thus, in order to have a convenient excuse for refusing the Count's proposal, Lavoisier and Marie-Anne quickly got married. The 'ruse' apparently worked, and the Count was put off.

The newlyweds, however, were devoted to each other for the rest of his life, and she proved to be a strong supporter and advocate for Antoine and of his career in particular.

Prior to his death, Madame Lavoisier had been continually at his side, able to 'talk shop' with Lavoisier on a routine basis. She translated journal articles from English to French for him to read. She participated in the actual experimental research as his laboratory assistant. She was even to elegantly produce many of the illustrative engravings featured so prominently in his publications, her illustrations forever archived within the important scientific literature in chemistry.

After his untimely death, Madame Lavoisier, who was spared the guillotine during the French Revolution, continued on with the work of her late husband. She made every effort to retrieve her husband's laboratory notes and books which had been confiscated during his prosecution. She was able to assure continued publications of later printings for his works, ensuring their permanent archival in the published literature. She also worked on and published his memoirs. She hosted many gatherings in which prominent investigators of the day were invited to present and discuss new scientific contributions in the newly established field of modern chemistry.

Madame Lavoisier lived for many years after the death of her beloved husband Antoine in 1794. Marie-Anne died on the 10th of February in the year 1836 at 78 years of age.

12) I think as a closing comment, I will quote Joseph Lagrange who said "It took but a moment to sever that head, though a hundred years perhaps will be unable to replace it." Lavoisier was known as the Father of Modern Chemistry? Has anyone replaced him?

I think that the quotation you to which you refer and which is rightly attributed to Lagrange, is a noticeably profound one, and I believe it is most certainly an *apropos* observation. Lavoisier was so prominent in the fledgling subject of modern chemistry with so many amazing discoveries it is difficult to comprehend what other additional great contributions he might have been able to make, had he not been quite unfairly executed at such an early age.

Many great scientists have made their greatest contributions to science in their later years. I believe that in terms of playing such an important role in the transition from alchemy, a dubious and defunct area, to the field of modern chemistry, a widely accepted and extremely important scientific discipline today, Lavoisier certainly has no equal.

Chapter 5

EDWARD JENNER

1) Dr. Varela, if there is one word that's used to strike fear into the hearts of many, the word would be smallpox. Can you first describe the devastating effects of smallpox, and the signs and symptoms of it?

Thank you very much for your interest in Dr. Jenner and this particular disease, smallpox. I think it is undoubtedly *apropos* that you invoked the term 'devastating.' Indeed, the smallpox disease has had not only profoundly devastating effects upon its individual victims, but has wreaked havoc with equally devastating effects upon countless human populations and was involved in several significant events throughout documented history.

Let's start first with your inquiry about the effects of smallpox upon the individual patient. The signs and symptoms would begin in the patient approximately 5 to 17 days after exposure to the so-called causative agent, i.e., the disease-causing microbe, which in this case is the Poxvirus variant known as *Variola major*. The initial sign of the disease included a high fever, plus symptoms of fatigue, backache, a terrible headache, and malaise (i.e., a feeling of discomfort and uneasiness). The next sign that developed in the patient was a rash in the mouth that spread to the body.

Almost immediately after the rash's appearance, the patient would experience vomiting, diarrhea and profuse bleeding from the rash. The characteristic rash of smallpox has also been referred to as vesicles, lesions, pustules, or more commonly, the pox. The misery associated with the smallpox disease would be alleviated in 15 to 40% of its victims by a merciful death. On the other hand, those who would survive the miserable disease were rewarded in the knowledge that they were immune to smallpox for the rest of their lifetimes. Such immunity to smallpox has been known since ancient times.

Now let's briefly examine the effects of smallpox in human populations. Smallpox virus is believed to have been the cause of several ancient epidemics, such as the so-called plague of Antoninus or the plague of Galen, after a Greek physician who reported on its effects at the time. Incidentally, Galen is famous for formulating the idea that disease was caused by imbalances in the four humors: blood, phlegm, yellow bile and black bile.

Another example, this one being somewhat notorious, involved the effects of smallpox upon the native inhabitants of the New Worlds and the newly discovered island continent of Australia. European explorers, who were immune, apparently unwittingly exposed these New World and Australian human populations to smallpox, with extraordinarily disastrous outcomes.

First, in terms of mortality, the losses were immense. These affected populations included, for instance, the Aztecs, the Incas, the Native Americans, the Aborigines, and the Hawaiians. In many of these cases, and others, the losses in the human population numbers provided distinctively strategic advantages to the new visitors, who were virtually immune to smallpox due to having been previously exposed over many generations, recovering and passing along their immunity to subsequent generations.

Thus, the combination of non-immune New World hosts and immune European guests depleted the host populations, essentially allowing the visitors to complete their conquests of the new lands. In addition to these horrific losses in life from smallpox, however, were also consequent losses in the cultures, traditions, knowledge, and even entire fledgling civilizations. Many of these affected populations never fully recovered and were even totally lost to history. More recently, it has been estimated that about 300 million people perished because of smallpox in the 20th century alone.

Lastly, let's consider several historical events that have involved smallpox. Probably the very first patient observed to exhibit clinical signs of smallpox was the pharaoh Ramses V, whose well-preserved mummified body (on display in a Cairo museum) clearly shows the pronounced and distinctive sign of smallpox: the pustules.

One prominent patient was George Washington, who is believed to have had a case of smallpox when he was only 19 years old. I know that in historical circles, one usually does not speculate what might have been. However, one can certainly imagine the history-changing consequences that may have occurred had the young Washington succumbed to the effects of smallpox and died at the age of 19. The American Revolution may certainly have been lost to the British, or even delayed, perhaps indefinitely. It is known, however, that Commander-in-chief Washington issued orders to have his entire Continental army vaccinated in 1776 against the smallpox during the war of independence against England.

Further, it is widely thought that Abraham Lincoln had a mild bout of the smallpox shortly after he delivered his famous Gettysburg address in 1863. Historians believe that Lincoln had passed the smallpox onto his son, Tad, who survived the smallpox, and onto his valet, William H. Johnson, who died from the disease.

Perhaps, however, the two most important figures in Dr. Jenner's time, at least as far as smallpox is concerned, are a young dairy milkmaid and a child. These two young people helped Dr. Jenner to change the course of history.

2) Edward Jenner, truly a giant in the field of science, developed a vaccine for smallpox. How did this come about?

Dr. Edward Jenner's development of his vaccine has an interesting history. Jenner was born on May 17, 1749 in rural Gloucestershire, England. He was apprenticed to a London-based surgeon and subsequently established a practice as a physician near his hometown in the English countryside, in an area notable for its dairy farms.

Jenner had been acutely aware of the ravages of smallpox in his patients. However, like Dr. Benjamin Jesty before him, Dr. Jenner similarly made the astute observation with his milkmaid patients who often had cowpox lesions on their hands from having had constant contact with cows they milked; that is, these milkmaid cowpox patients never seemed to get the smallpox.

In 1789 or 1790, using infected pus material from cowpox infected pigs, Dr. Jenner inoculated his 10-month old son. What he did next was both novel and startling; he repeatedly injected his son with smallpox! Fortunately, his son never acquired the smallpox, indicating that he was immune to smallpox, perhaps from the cowpox inoculation. Interestingly, this little experiment per se did not alter scientific history, as Dr. Jenner did not publish it.

Instead, Dr. Jenner's famous history-changing experiment, published in 1798, involves another experiment, this time taking cowpox lesions from the infected hands of dairy milkmaid, Sarah Nelmes (whose cowpox-laden hands are illustrated in Jenner's publication), and using her infected material to inoculate the 8-year old James Phipps by injection. Phipps then acquired a mild case of the cowpox. What Dr. Jenner did next was equally as startling as he had done before with his son; he then injected Phipps with lethal doses of smallpox! The boy Phipps survived. Next, Dr. Jenner repeated this same procedure with 16 additional people, recording each of their names in his publication and describing their outcomes in detail.

Thus, Dr. Jenner provided the first experimental evidence validating the effectiveness of using cowpox as a vaccine for smallpox. The term vaccine or vaccination originates from the Latin term *vacca*, meaning cow. The vaccination procedure provided the advantage over variolation (using smallpox lesions to inoculate against the smallpox) in that cowpox infections were relatively mild and did not result in accidental smallpox deaths, as seen with smallpox material used with variolation.

3) What other discoveries did he make during his career? And was he acknowledged for his contributions during his lifetime?

It is interesting that Dr. Jenner, a small town surgeon, was also an established field naturalist. In fact, he was noted for a discovery having nothing to do with smallpox. In 1787, Dr. Jenner studied birds, namely, the cuckoos, in which he observed that newly hatched cuckoos in their nests tended to remove the eggs of their siblings from their nests. Prior to this, it was widely believed that the adult cuckoos were responsible for the egg removal from their nests. This discovery led to Dr. Jenner's election to the Royal Society two years later.

Dr. Jenner's contributions with vaccinations using cowpox material against smallpox, however, was met with immediate acceptance. Consequently, Dr. Jenner was lavishly celebrated with cash awards, such as 10,000 pounds from the thankful House of Commons, and numerous honors, including one from Napoleon himself. Even the famous Prof. Pasteur honored Dr. Jenner some 80 years after his death by coining the term vaccination, to commemorate Jenner and his work with the cowpox and the smallpox.

4) In terms of other scientific discoveries at that time- what was happening in this field of "vaccinations"? And is there any one single person responsible for this conceptualization?

The story of smallpox leading ultimately to the development of vaccinations involves many individuals and has its starting points in ancient China and India, in which pus from smallpox lesions was collected with a thread or a needle and rubbed into a scratch or a puncture on the skin, providing protection from the smallpox. The vaccination technique and its variations have also been called inoculation, engrafting, variolation, and immunization. Based on these Chinese and Indian procedures, Giacomo Pylarini conveyed in 1701 that he took dried crusts from smallpox lesions and used them to inoculate children in Constantinople, reporting good success.

Hearing of this account in 1721, Lady Mary Wortley Montagu, the wife of the British ambassador to Constantinople, introduced her children, and later England, to the technique, which grew in popularity and became widespread throughout Europe. During the same year, at the urging of pastor Cotton Mather, Dr. Zabdiel Boylston of Boston invoked the technique to quell a smallpox epidemic but with mixed success, as several of the inoculated patients later actually acquired the smallpox and died from it. Both Mather and Boylston were blamed for the deaths and attacked, both in publications and in person, being physically assaulted, in fact.

In 1764, Angelo Gatti reported that attenuation of the smallpox virus prior to inoculation might make the technique safer. Then, in 1774, Benjamin Jesty was the first to use material from cowpox lesions to inoculate against smallpox in humans, using his wife, son and himself as recipients of the cowpox pus from infected cow udders. The Jesty family shortly afterwards survived a smallpox epidemic in England. This was essentially the state of the knowledge in Jenner's day prior to his experimentation and set the stage for his work with cowpox and smallpox.

5) Many scientists died not knowing the impact of their work. How pervasive is the concept of "vaccination" today and how impactful was Edward Jenner's work?

Dr. Jenner's work resulted in a profound impact on world history. After his experiments were published, vaccination became widespread throughout the world. And although there were pockets of resistance to vaccination, due mainly on religious grounds, the success of Jenner's vaccination work cannot be overstated. First, the vaccination principle was employed against many other infectious diseases, such as the dread rabies by Prof. Pasteur, plus chicken pox, measles, mumps, rubella, yellow fever, anthrax, cholera, diphtheria, Lyme disease, pertussis, tetanus, tuberculosis, typhoid fever, and polio.

Second, because of Dr. Jenner's work, it led ultimately to the total eradication of the dread smallpox disease from the face of the Earth, with the last known natural case of smallpox being officially recorded in 1977, in a patient named Ali Maalin. Thus, smallpox was the first (and only) disease in history to be totally eliminated through the combined concerted efforts of vaccine research and extensive vaccination programs.

6) Is there anything I have neglected to ask?

I think it may be interesting to note that even though smallpox is the only disease to be abolished through the far-reaching efforts of humankind, polio is, as of this writing, slated to be the next disease for which global eradication is just around the corner. I think it's amazing that such disease eradication has been realized in our lifetime, all due to the experiments of many but with the notable ones conducted by Dr. Jenner. One can certainly envisage a time when perhaps all infectious disease shall be history.

Chapter 6

IGNAZ SEMMELWEIS

1) Professor Varela, I see it in almost every bathroom I go into. There is a sign "Have You Washed Your Hands Today." And obviously on every doctor show on T.V. we see those E.R. docs scrubbing their hands away. Let's give credit where credit is due- Was it Ignaz Semmelweis in the 1850's who started all this handwashing procedures?

You are absolutely correct! It was indeed Dr. Ignaz Philipp Semmelweis who began the early push aimed at invoking proper handwashing procedures. It was a practice that he had intended for medical students and their physician teachers. His efforts started the modern antisepsis and disinfection practices.

Today, there are periodic handwashing campaigns, encouraging people to wash their hands. In these campaigns, a variety of ways are used to encourage others to wash their hands, e.g., in bathrooms as depicted on the signs you've encountered. Handwashing prevents infectious disease transmission through populations.

2) Where exactly was Ignaz Semmelweis born and when? He sounds Swiss to me!

"Ignác Fülöp Semmelweis" was born on the first of July, 1818, in Budapest, Hungary. His German-speaking mother, Teresa Müller, was the daughter of a prosperous Bavarian manufacturer of horse-driven coaches, and Ignaz's father, Joseph, was of German descent and wealthy vendor of spices and grocery goods. Their official citizenships were in the Kingdom of Hungary.

3) What were his early years like and where did he get his education?

The young Semmelweis spoke German and did not learn Hungarian until he attended secondary school, resulting in a distinctive dialect. During this time, he apparently developed a strong dislike for writing. Together, his accent and disdain for writing would, in part, have profound ramifications for Semmelweis and his career later in life.

Semmelweis attended university in Vienna to study law in 1837, but became disinterested in the topic after having attended a life-changing lecture by Prof. Joseph Berres dealing with the subject of gross anatomy. Shortly afterwards, Semmelweis switched to the study of medicine, first at Vienna and briefly at the University of Pest, and then returning to Vienna to finish his doctorate in medicine in 1844.

4) Is it known whether his "handwashing" practices were the result of some science experiment, or observation or did he learn this from other older, wiser doctors?

In a word, Semmelweis based his handwashing policy on observation. As a newly minted doctor working in a maternity ward at the General Hospital in Vienna in 1846, Semmelweis became concerned with the high rate of mortality in women patients while giving birth. The primary cause of death was the so-called childbed fever, or puerperal fever, also known as puerperal sepsis. Semmelweis started collecting data in two distinct maternity clinics. One of these clinics was run mainly by female midwives, and the other clinic was run primarily by male doctors and their students of medicine. Semmelweis observed that the puerperal fever mortality rate in the doctor and medical student clinic was significantly high (between 10 and 30% mortality) compared to the midwifery clinic (only 1 to 3% mortality). The reason for this difference in the death rates was unknown at the time.

What was clearly known, however, was the stark difference in death rate between the two wards. The community certainly knew, and women who were in labor tried their best to avoid the doctor ward, often prolonging the wait to be admitted just so they could be sent to the midwives. Women who were admitted to the doctors' section would be dismayed and rightfully frightened. Semmelweis himself recorded how women in labor and in the doctors' ward would kneel down on the floor and beg the doctors to be discharged so they could be free to go to the midwives.

At first glance into the problem, Semmelweis noticed that in the clinic staffed by the midwives, the mothers giving birth did so while laying on their sides while in the doctor clinic, the mothers laid on their backs. So Dr. Semmelweis instructed the doctors to tell their patients to lay on their sides, too. Unfortunately, the mortality rates due to puerperal fever remained unchanged in the male doctor clinic—the mortality rates stayed high.

Next, Semmelweis turned to an observation he made regarding the doctor and medical student clinic; each time a patient died of the dread childbed fever, a priest would slowly and somberly walk through the maternity wards while a hand bell was rung, signifying that a death had occurred in the hospital. Semmelweis reasoned that this priestly behavior had so greatly terrified the mother patients that they literally took sick and were consequently prone to death by the puerperal fever. Thus, Semmelweis instructed the priests to stay clear of the ward during their walk-throughs and to lose the bell ringing while they were at it. Unfortunately, this policy change had no effect on the puerperal fever death rates, either.

Then, an incident occurred which inspired Semmelweis to invoke a new notion to explain the high number of deaths in the doctors' ward. The incident was the death of his good friend and colleague, Prof. Jakob Kolletschka, a medical doctor from the doctor ward. Like all of the doctors and medical students of the doctor clinic, Dr. Kolletschka had been routinely conducting autopsies on cadavers and had accidently cut his finger with a scalpel while working on a mother who had perished from the childbed fever. The doctor himself soon then got the childbed fever and shortly thereafter died from it.

While the concept had been known for some time as the so-called "miasma" or bad air, in which disease was due to something foul in the foul-smelling air, Semmelweis reasoned that the mortality rate difference between the wards may be due to "cadaver particles." The doctors from the ward were also working on cadavers and may have unwittingly carried these cadaver particles to their women patients in labor, giving them the puerperal fever!

So Semmelweis instructed his doctors and their medical students to start washing their hands in a bowl containing a bleach type of solution between autopsies and before their visits with maternity ward patients in the doctor ward. It was hoped that the handwashing practice would eliminate or reduce the cadaver particles from the doctors and medical students and thereby prevent transmission to the women patients. After invoking the handwashing practice, mortality rates of the childbed fever subsided within the doctors' ward. It was a dramatic result.

5) How long did it take before his practice caught on? And did he receive the recognition he deserves?

As a first approximation one might have reasonably concluded that doctor handwashing between patients, especially if backed up by the carefully gathered data of Semmelweis, would have immediately caused the practice to become widespread throughout all or most hospitals worldwide. Unfortunately, this was not the case at first, even surprisingly not the case in Semmelweis's own hospital. From the very beginning certain events worked against Semmelweis to confound his efforts to implement a permanent policy of handwashing for the doctors and the medical students.

Initially, and by coincidence, a new ventilation system had just been installed in the hospital by his director, Dr. Johann Klein, during this same timeframe as the handwashing practice had begun, and the decreased mortality rate was mistakenly attributed to the better ventilation, rather than the handwashing.

Next, historians postulate that a combination of three seemingly disparate obstacles worked together further to confound Semmelweis at his hospital. The first obstacle was the refusal of the older generation to institute new practices, like handwashing, and the second obstacle was the hospital bureaucratic red-tape. The third obstacle, however, may have been Semmelweis, himself. It was reported that Semmelweis became brash, often lacking in tact when he disagreed with others, resulting in the emergence of new enemies, including his own director, Prof. Klein. The doctors and medical students eventually gave up the handwashing practice and Semmelweis was fired in 1849. Perhaps feeling insulted by a less-than-adequate re-hire, Semmelweis left Vienna five days after starting the new post, without an explanation to anyone, even to his friends. The exact reason for his abrupt departure is still a mystery to this day.

Another important factor to consider is the fact that Semmelweis was not able to provide a definitive explanation why handwashing worked to curb the puerperal fever. This was not entirely his fault, however, as no one at the time yet knew about the germ theory of disease that microbes could cause infectious disease, a theory purported by the great Prof. Louis Pasteur in the 1850s and experimentally demonstrated by Prof. Robert Koch in the 1880s. Thus, Semmelweis did not know his handwashing practices simply washed away the sepsis-causing microbes from the hands of the doctors and medical students. His cadaver particles were bacteria, but no one knew then. Opponents of handwashing ridiculed the notion that tiny cadaver particles could be so powerful as to kill a much larger human being. His fellow doctors were further put out by the idea that they themselves could be the transmitters of the puerperal sepsis. They found that idea abhorrent; and it was more convenient to invoke the idea that the sepsis contagion was brought to their hospitals by the women patients from the community, rather than to the patients from the doctors and their cadavers.

Further, Semmelweis refused to publish his findings; you'll recall his early disdain for writing. Additionally, he was periodically teased by his adult colleagues regarding his accent. It is speculated that Semmelweis was consequently insecure about his writing and professional work. Whatever the case may have been, he simply did not publish his handwashing ideas. This was a damaging final blow for any immediate and long-lasting implementation of handwashing by the doctors of Vienna.

After his dismissal and his abrupt departure, he took an unpaid post at the St. Rochus Hospital back in his native Hungary, in Budapest, where Semmelweis continued his handwashing practices, improving the survival rates for their patients. Unfortunately, his quarrelsome and unpleasant personality resurfaced there and again worked against him. Hence, he left that post to start a private practice and later took a new position as a professor at the University of Pest, in 1855.

Semmelweis would occasionally read criticism of his handwashing methods in the medical journals, prompting him to write to sympathetic and seemingly like-minded maternity doctors, hoping to win their public support and help to vindicate him, but to no avail. He was largely ignored.

Therefore, Semmelweis overcame his aversion to writing long enough to write and publish a book in 1861, discussing his arguments in favor of handwashing and hoping to be finally vindicated. His book, however, was longwinded, convoluted and difficult to understand. To make matters worse, he publically insulted critics of his book and ideas with name calling. Not surprisingly, therefore, even the publishing of his theory did not result in widespread adoption of handwashing by doctors or of their students.

Unfortunately, the handwashing practice did not become widespread in surgical wards until several years after the death Semmelweis, in the 1870s, after Prof. Louis Pasteur formulated his germ theory of disease and with Dr. Joseph Lister, who first instructed his surgical personnel to disinfect the surgical wards.

Sadly, the situation with Semmelweis's personal life never managed to improve. His behavior in later years turned out to be increasingly short-tempered, unpredictable, and unpleasant. He became forgetful, occasionally not knowing where he was. His nervous breakdown complete, Semmelweis was committed to an insane asylum. While institutionalized he was reportedly mistreated and beaten so severely that he developed a chest infection as a result of the trauma he received from being beaten. He soon died at the age of 47 on August 13, 1865, from the sepsis—the very same disease he tried to wash away. Consequently, Semmelweis never received the recognition he deserved in his lifetime.

6) What were some of his other contributions or discoveries?

One of Semmelweis's greatest contributions to science and medicine was his notion that simple handwashing by doctors using a bleach-containing solution in a bowl could not only prevent the onset of the dread puerperal sepsis, he actually came close to realizing the germ theory. He did correctly conclude that his cadaver particles, now known as bacteria, could be transmitted to others by contaminated hands.

It is speculated that had he used a microscope, he might have been able to show that his cadaver particles were in fact microbes, and he would have put a severe dent in the miasma theory (foul air) and in the doctors' inaccurate use of contagion theory to argue that the disease was brought into the hospital by the women patients, themselves, from the community. But this did not happen for Semmelweis.

Semmelweis unwittingly made another contribution, the so-called 'Semmelweis Reflex.' The 'Semmelweis Reflex' is defined as the rapid and total rejection of novel ideas or practices, because they go against firmly established practices or beliefs, even in the face of supporting empirical data. This reflex phenomenon was named after Semmelweis by those who formulated it.

7) For the average citizen, it seems common sensical, but why should the average student in school, or worker in a fast food establishment or restaurant wash their hands?

Semmelweis's work led to the later discovery that hands contaminated with disease-causing microbes are common and significant transmitters of infectious disease. This is because people frequently use their hands to touch many objects around them and other people. Objects that are contaminated by disease-causing microbes may be agents of transmission to others.

Another common mode of infectious disease transmission involves food or drink contaminated with pathogenic microbes. Often, these disease transmissions encompass the so-called "fecal-oral" route, in which the patient consumes fecal matter present in the food or drink.

Therefore, it is critically important that students in schools, in constant contact with many others, and food preparers, wash their hands in order to avoid microbial disease transmission and spread. In general it is important to wash your hands (warm water and soap for at least 20 seconds) before, during, and after preparing meals.

It is also important to wash hands (a) before consuming meals; (b) before and after providing healthcare to patients; (c) after using the bathroom or toilet; (d) after changing a diaper; (e) after handling waste, like garbage or feces; (f) after contact with animals or their feces; and (g) after coughing or sneezing into your hands, or blowing your nose. These practices help reduce the conditions that foster movement of pathogenic microbes through populations.

Alternatively, a common practice is to cough and sneeze into one's sleeve on the inside of one's elbow. This reduces hand contamination with pathogenic microbes.

On the "other hand," as important as it is to wash hands for the reasons listed above, it's also important to avoid getting carried away with it and acquiring the obsessive-compulsive behavior disorder of constantly and needlessly washing your hands throughout the day and night. The abnormally excessive handwashing may also damage the skin and cause the hands to become prone to cracking and bleeding. It may also wash away the healthy bacteria that reside on the skin (skin microbiome) which protects people from the rare pathogenic microbes.

8) What have I neglected to ask?

I think it's interesting to note that necessary handwashing is currently not a practice that's followed by the great majority of humans. Even healthcare workers are not washing hands between patients 100% of the time. In fact, in intensive care units of healthcare facilities handwashing was reported to range from only 26% to 50% of the time, with the latter higher figure resulting only if monitoring systems are present.

This is a serious issue because many of the truly dangerous bacterial superbugs, those with multidrug resistance and which are becoming increasingly difficult to treat effectively, are emerging from healthcare facilities and moving to neighboring communities. Some estimates state that, due to inadequate handwashing between patients, the morbidity rates number just over a million cases annually while the mortality numbers are in the hundreds of thousands per year. Healthcare facility acquired infections (HAI) have emerged as a serious public health concern on a worldwide scale. Lack of dedicated handwashing may be a primary cause of HAI; washing hands between patients is a fundamental solution. So the question that you have not neglected to ask about, thankfully, is "Have you washed your hands today?"

Chapter 7

GREGOR MENDEL

1) Dr. Varela - if there is one name that stands out from high school biology, it is Gregor Mendel, and his work with dominant and recessive genes. Let's get some background. Where and when was he born and what did he first start out as?

The Father of Genetics, Johann Gregor Mendel, was born on the 22nd day in the month of June in 1822 in Heinzendorf, a village now called Hyncice, which was located within the province of Silesia, in the Hapsburg Empire of Austria, and now currently an enclave of the Czech Republic (Czechoslovakia).

The name given to Mendel by his parents on the day of his birth was Johann, and Gregor was the name he took when he became ordained as a monk in 1851. Mendel's father was Anton, a peasant farmer, and his mother was Rosine Schwirtlich. Only two of Mendel's siblings survived early childhood, an older sister, Veronica (b. 1820), and a younger sister, Theresa (b. 1829). The Mendel family lived in poverty during his childhood.

In 1840, Mendel graduated from the gymnasium in Opava (also referred to as Troppau) and entered the Olomouc Philosophical Institute, which was part of the University of Olomouc (Olmütz), and in 1843 he graduated from this institution. Shortly thereafter, in 1843, Mendel was admitted to the St. Thomas monastery housed in Brünn (now called Brno) and became a monk.

2) Now, the big question in science is often WHY-? Why did he get involved in science, WHY did he get involved with genetics, and WHY peas?

There is an interesting story regarding Mendel's choice to study the sciences while in the monastery. Apparently, his choice of studying science was, at first, dictated by a failure.

According to the story, in 1851 Mendel failed the natural science portion of his qualifying examinations for his teaching license; therefore, Mendel was sent to the University of Vienna in order to become proficient in the natural sciences.

While at university, Mendel had enrolled in courses pertaining to systematics of botany, physiology, microscopy, morphology, and mathematics, becoming quite proficient in this particular area. I believe that the genius of Mendel was his unprecedented ability to combine the field of mathematics to that of heredity, in order to become the father of genetics.

Additionally, Mendel became accustomed with plant and animal breeding knowledge and practices. Thus, Mendel had now acquired an expertise in agriculture and botany, alongside his foundation in theology. Mendel had been aware of the studies by Joseph Gottlieb Kölreuter in 1761 showing the formation of so-called plant hybrids. It is this work that had so interested Mendel.

Upon his graduation from university in 1853, Mendel returned to his monastery in order to teach, and he began gardening in order to pursue his interests in studying the hereditary nature of plants and their hybrid offspring.

In particular, Mendel studied *Pisum sativum*, a variety of sweet pea plants, for several important reasons. First, the pea plants generated rather large numbers of seeds, in which he could uniquely apply his knowledge and expertise in mathematics towards analysis of these seed numbers. Further, Mendel noticed that his pea plants had relatively shorter generation times, compared to those of, let's say, animals and other plants. Thirdly, Mendel found that his pea plants had easily observable and distinctive characteristics, like the shapes of the seeds (round peas versus wrinkled peas) or of the seed pods (swollen versus thin), the colors of the seeds (green or yellow), the colors of the flowers (white or red), the locations of the flowers along the stem (proximal versus distal), and the lengths of the stems (tall or short).

Mendel used these easily observable plant traits and his horticultural pollination techniques to produce so-called pure-bred pea plants, i.e., plants which repeatedly produced offspring harboring the same traits during each subsequent generation (e.g., offspring always produced round peas or always wrinkled peas).

These efforts took Mendel two years to accomplish—it was a tremendous effort on his part. But he finally acquired his needed and pure lines of pea plants, each line with their consistently generated individual pure traits. These pure plants have been referred to as the so-called parental (P) generation.

He was, thus, ready to conduct his hereditary experiments on painstakingly generated pure parental pea plants, generation P.

To begin with, Mendel cross-pollinated his various generation P plants with each other. For example he crossed-bred the tall plants with the short plants, or his green pea plants with his yellow pea plants, or his round pea plants with his wrinkled pea plants, or his white flowered plants with his red flowered pea plants, and so on.

In order to do his cross-pollination experiments using his purebred pea plants, Mendel performed the following procedures to do so.

First, from one pure plant (e.g., a tall one) Mendel removed the stamens, which contained the pollen and brushed the stigmas of the other pure plant (i.e., a short pea plant) and allowed the cross-pollinated plants to produce flowers, which then produced pea pods, containing the seeds, i.e., peas. Mendel then planted the newly emerged pea seeds and made observations of the character traits of his new progeny plants, which are referred to as the F_1 generation, for the "first filial" generation.

Initially, Mendel reasoned that the F_1 generation would be a mixture of the two pure traits, e.g., medium height, if the cross-bred P generation individuals were tall and short plants.

To his great surprise, Mendel found that certain traits disappeared completely! For instance, his F_1 generation plants were all tall; the short trait failed to show itself in any of the F_1 generation progeny.

The main question that was burning in Mendel's mind was: what had happened to these "disappeared" traits? Did they go away permanently—never to be seen again in later generations? Or, were they dominated by the traits that had manifested themselves in all of the progeny of the F_1 generation?

In order to provide answers to these perplexing "disappearing traits" results, Mendel conducted another set of experiments. Here, he cross-bred individual plants from the F_1 generation to other F_1 generation plants, to make a third generation, denoted as an F_2 generation; that is to say, he performed an F_1 x F_1 cross, thus, producing an F_2 generation of pea plants. In these F_2 generation pea plants the repressed traits re-emerged, but only in a certain subset of them, certainly not all of them.

In other words, the ratio of the dominant traits to the suppressed traits in the F_2 generation was roughly about 3:1 for each of the various traits. Since the traits that were repressed in the F_1 generation had not disappeared forever and had actually reappeared in the F_2 generation, Mendel concluded that the traits had been prevented from manifesting themselves by the dominant traits. The traits that had been suppressed were termed recessive.

Thus, Mendel had discovered the principles of dominance versus recessiveness. The first principle in the field of genetics, dominance, had now been discovered, and it was Mendel who was the very first person to discover it.

3) Back in his day, there was no internet or web- how did he communicate his scientific results to others?

Although Mendel was ordained as a monk, with its required education in theology, he was also trained, nonetheless, as a *bona fide* scientific investigator, complete with practical knowledge in the methods he needed to conduct his experiments and in the general scientific method for expanding the frontiers of new scientific knowledge.

Thus, like any scientific experimentalist of his day, Mendel shared his original data in a public presentation venue and in a permanent publishing of his work in print.

There is an accounting, as told by biographers, that Mendel, in an effort to spread his knowledge about, had sent a reprint of his 1866 paper to other scientific investigators, whom Mendel thought might perhaps be interested in the novel nature of his work, including, it is said, a rather famous investigator, Carl von Nägeli (or Karl von Näegeli), an eminent botanist.

Dr. von Nägeli had previously conducted studies of cellular division, pollination, gamete production in plants, and especially in hawkweed plant taxonomy. In his accompanying letter to von Nägeli, Mendel also included some of his pea seeds, requested some hawkweed seeds, a favorite plant of von Nägeli's, and then offered to test his inheritance theories in the hawkweed plants.

At first, von Nägeli ignored Mendel's correspondence for what seemed like to Mendel an unusually long period of time of silence. When von Nägeli finally replied by letter, he criticized Mendel's findings as too preliminary to make any definitive conclusions regarding heredity—keep in mind that Mendel had cultivated an estimated 10,000 pea plants in order to come up with his correct inheritance concepts.

It has been reported that von Nägeli was leery of Darwin's theory of evolution. Consequently, von Nägeli had trouble in believing Mendel's inheritance ideas. In his same correspondence to Mendel, von Nägeli included the hawkweed seeds that Mendel had requested. Incidentally, there is no evidence that von Nägeli did anything with Mendel's pea seeds.

In contrast, Mendel duly cultivated von Nägeli's hawkweed plants, and the data became obfuscated. Unbeknownst to both Mendel and von Nägeli, or to anyone else at the time for that matter, individual hawkweed plants each have inseparable allelic variants and unusual gene clusters; plus they mate both sexually and non-sexually. Thus, the Mendelian approach to the study of hawkweed inheritance did not apply. The work went nowhere.

One historian has opined that von Nägeli's advice to Mendel ("study the hawkweeds") was the worst advice any investigator ever gave to another investigator. Indeed, had Mendel studied the hawkweed first, it might have set back the founding of genetics as a *bona fide* scientific discipline for generations of scientists to come.

4) What was the immediate feedback about his discoveries or theories?

Mendel's very first historical release of his heredity work came about in February of 1865 during a scientific conference in which he presented his data to a nearby local group, called the Brünn Society for the Study of Natural Sciences. During this very first presentation of his work, Mendel took his audience, who were primarily botanists, through much of his work, describing his methods, his mathematical data, and his conclusions about dominance, independent assortment, and independent segregation.

To Mendel's disappointment, no one asked questions afterwards, even though it was (and still is) customary to ask questions at scientific seminars, conferences, colloquia, and symposia. Biographers of Mendel speculate that no one in the audience of early botanists understood his mathematics-based presentation, presumably due to Mendel's emphasis on the use of numbers to sort out inheritance patterns of pea plant traits.

A second presentation was given by Mendel to the same Society only a month later, this time expanding on how he had used his mathematical methods to arrive at his genotypic and phenotypic ratios and to arrive at his notions of inheritance. Once again, his audience sat in stunned silence—apparently in a complete incomprehension of the novel and unprecedented genetic principles just presented by Mendel.

It was, unfortunately, virtually incomprehensible to the members of the Society how inheritance patterns of clearly visible traits in plants could be arrived at simply by using numerical data and analysis. At this point, it is recorded that several members of the Society even felt sorry for poor old Mendel, for having wasted so many years (about 7 years) of his life conducting painstaking cross-breeding work and "counting peas" only to be met with virtual incomprehension, if not outright disbelief.

Nevertheless, in 1866, the Society published his presentations in its dedicated scientific journal, called the *Proceedings of the Natural History Society of Brünn*. It was written in German. The seminal paper was duly bound and sent by the society publisher to the main libraries worldwide, where it was summarily ignored and collected dust in the libraries' journal stacks for more than 35 years.

5) What else did he investigate and what else did he discover?

As I mentioned above, Mendel formulated the genetic principles of dominance versus recessiveness with respect to characteristic traits in living beings like his pea plants. A dominant trait would somehow mask a corresponding recessive trait. From his findings, Mendel formulated early on other important principles of genetics.

He reasoned that because the various recessive pea plant traits disappeared for a generation and then actually came back in the next generation but at certain mathematical ratios, the traits must therefore be determined by so-called discrete factors that he called "unit characters." We now know that these so-called unit characters are what we modern investigators call genes.

Mendel further reasoned that these unit character genes occurred in pairs: one unit character (or gene) originating from one of the parents and the second unit character gene arising from the other parent.

Another concept arising out of Mendel's studies is referred to as the so-called segregation of alleles. An allele is a term given to an alternative form of a particular unit character, or gene. For example, one allele of a gene for flower color may dictate red flowers and the other allele of that gene for flower color may specify white flowers, in the pea plants.

In any case, Mendel's principle of segregation states that an individual's set of two versions (i.e., alleles) of a gene will separate from each other (i.e., segregate) during their transmission to the next generation, like from a parental generation to an offspring F_1 generation. That is to say, when gametes or sex cells from an individual form, like an egg or a sperm, the alleles separate into each of these types of formed gametes.

Since the gametes have only one of the gene's two allelic versions, they are termed haploid. Whereas, on the other hand, the individual organisms with their full complement of gene versions, i.e., both alleles of the gene are present, are referred to as diploid.

The resulting offspring may inherit one allele from one parent and the other allele from the other parent. In 1905, Reginald Punnett invoked the use of the so-called Punnett Square (See Figure 1) to distinguish between the genetic natures of the parents versus their gametes versus the genetic nature of their offspring.

First parent's gametes: Second parent's gametes:	R haploid (red)	r haploid (white)
R haploid (red)	RR diploid (red)	Rr diploid (red)
r haploid (white)	Rr diploid (red)	rr diploid (white)

Figure 1. An F_1 x F_1 hybrid cross between parental individuals with Rr genotypes.

In Figure 1, in which, for example, the flower color trait is considered, R is the allele for the gene that specifies flower color, red, which is dominant, and r represents the allele of the flower color gene that specifies a white flower color and is recessive.

Each parent produces two gamete types: R (dominant, red flower color) and r (recessive, white flower color). There are three types of outcomes in the offspring: homozygous dominant (RR), homozygous recessive (rr), and heterozygous dominant (Rr).

The genotypes of the offspring produced in Mendel's work occurred as a ratio of 1: 2: 1 for the RR: Rr: rr, and a flower color trait as a phenotype (i.e., an observable trait) in a ratio of 3: 1 for red: white.

As an aside, in Mendel's publication he referred to these red flower colors as being red-purple.

In other work, Mendel then developed the methodological concept of the so-called "test-cross" analysis, in order to evaluate the genotype status of an individual plant. That is, Mendel wanted to determine whether an offspring, let's say with a red flowered pea plant, was either genotype RR or genotype Rr.

The test-cross procedure involved crossing the yet unknown RR or the Rr individual with a known rr, a homozygous recessive, individual.

If the genotype of the individual was RR, then the resulting new offspring in a test-cross with an rr individual (Figure 2) would be in a 4: 0 ratio of red versus white—all offspring would be Rr—hence, all offspring will have red flowers.

There are no RR or rr offspring in the test-cross of an RR individual with an rr individual, and these types of progeny result are, therefore, not observed in the Punnett square; and, as can be seen in Figure 2, only Rr individuals are observed in the progeny.

Homozygous recessive's gametes, rr, for test-cross: Individual's gametes:	r haploid (white)	r haploid (white)
R haploid (red)	Rr diploid (red)	Rr diploid (red)
R haploid (red)	Rr diploid (red)	Rr diploid (red)

Figure 2. Test-cross in which an individual has an RR homozygous dominant genotype.

If, on the other hand, the genotype of an individual was Rr, then the resulting new offspring in a test-cross with rr pea plants, see Figure 3, would be expressed in phenotype ratio of 1: 1 of red flower color versus a white flower color—half of the offspring would be red (Rr) and the other half would be white (rr). There would be no RR in this latter test-cross progeny.

Homozygous recessive's gametes, rr, for test-cross: Individual's gametes:	r haploid (white)	r haploid (white)
R haploid (red)	Rr diploid (red)	Rr diploid (red)
r haploid (white)	rr diploid (white)	rr diploid (white)

Figure 3. Test-cross in which an individual has an Rr heterozygous dominant genotype.

Another concept that was developed by Mendel was the notion of independent assortment. During this new concept formulation, Mendel speculated whether, on the one hand, any two given unit character genes would always be linked to each other—this meant that during gamete formation, the two linked genes always stayed in association with each other. On the other hand, the two unit character genes could very well assort independently and randomly of each other.

Mendel put these two distinct hypotheses, i.e., linked versus independent assortments, to the test, using a so-called two-factor cross breeding experiment in which his parental pea plants each had two pure traits, and cross-pollinated these parental individuals, producing new F_1 generation plants.

Next, he crossed individuals from the F_1 generation plants with each other to produce new F_2 generation plants. These F_2 generation progeny had phenotypes that were mathematically consistent with Mendel's notion of independent allelic assortment during the formation of gametes.

I think it is an interesting fact that Mendel not only studied pea plants, for which he is now quite famous, he studied string beans (scientific name is *Phaseolus vulgaris*) and corn (scientific name *Zea mays*), as well. Many of his inheritance principles held up to his own scrutiny in these plants just as well as they had for his pea plants. In this regard, Mendel is said to have been quite lucky in choice of organisms for study, in addition to being an astute scientific research investigator.

6) How was he received at the time, was his work acknowledged?

Aside from Mendel's 1866 paper having been cited only 3 known times in the intervening years between its initial publication and its later re-discovery by the established scientific world in the early 1900s, Mendel died in virtual obscurity, never knowing what a great contribution of such historical and scientific magnitude it was to become. It is said that after he died, on the 6th of January, in 1884, at 61 years of age, all of his papers were burned, sadly, in the very same yard where his pea plant garden had been located.

Mendel's scientific work had not been uncovered from its obscurity until 16 years after his death. The number of investigators with claims to having independently re-discovered Mendel's work, increases with time. I will mention a few notable examples here.

One of the best known of these so-called re-discoverers of Mendelian genetics is Dr. Hugo de Vries, a noted plant physiologist and professor of botany at the University of Amsterdam. Between 1892 and 1896, years after the death of Mendel, in 1884, de Vries had studied the breeding characteristics of the evening primrose plants and actually obtained the famous 3: 1 offspring ratios that Mendel had beforehand obtained. While waiting for an opportunity to publish his works in a monograph, de Vries had serendipitously stumbled upon Mendel's published works, and he (de Vries) then consequently published his own findings, providing a statement of confirmation for Mendel's work alongside his own, starting in 1901. Prof. de Vries was also one of the first to propose the mutational nature of the gene.

Another independent confirmation of Mendel's work was brought about by Karl Erich Correns, a German botanist, who also studied pea plants and who also acquired quite similar phenotypic ratios as Mendel and de Vries had obtained. When Correns sat down to write up his manuscript detailing his findings, he also encountered Mendel's seminal paper; and Correns published his work in 1900, providing evidence in support of Mendel's results pertaining to botanical inheritance.

Later, Correns was to follow-up on the fledgling genetics field and build upon the Mendelian principles of inheritance. For instance, Correns had discovered that genes could be physically linked to each other along their chromosomal structures, a finding that was later to be beneficial towards mapping the actual positions of genes located within their genomes. Interestingly, he found that other plants passed along their genetic complement onto succeeding generations via non-Mendelian inheritance patterns.

Lastly, Correns was the first investigator to correlate Mendel's so-called law of segregation to the process of meiosis, during which paired genes undergo separation of their alleles. This was an important discovery. Without Mendel's earlier supporting work to provide a context, later advances in our understanding of genetics might have very well been even further delayed.

Another line of supporting evidence for Mendel's work arose in 1900 from the bacteriological studies by Martinus Beijerinck, who found that the bright red color, produced by the bacterium named *Bacillus prodigiosus*, also called *Serratia marcescens*, could be altered to produce no color or color variations. Prof. Beijerinck interpreted these microbial phenotypic findings in terms of genetics and gene mutation. This work started a new line of research delving into microbial genetics.

In 1902, Prof. Rollins Adams Emerson, an American and plant geneticist, had confirmed Mendel's work with beans with his own similar type of work with bean plants, incorporating Mendel's concepts with his own and publishing this work. Emerson later follow-up on the bean work with genetic studies of maize (corn), a plant which was later to prove decisive in the later discovery of the so-called "jumping genes" by Nobel Laureate Barbara McClintock, a phenomenon known as DNA transposition.

7) His later years - what did he do in the twilight of his career?

Unfortunately, in 1868 Mendel discontinued his ground-breaking inheritance studies in order to become an administrator, the abbot of the Brno monastery, and his twilight years were spent partaking in unending managerial tasks and, unfortunately, internal and external politics. Along with his monastic administrative duties, Mendel became a curator and chairperson to a variety of local committees; he also stood firm against newly imposed taxes, which Mendel had believed were unfair, especially to the poor.

Unfortunately, an opposing faction within his own monastery had tried unsuccessfully to imply, if not outright claim, that he had gone mad. Nevertheless, the tax fight took out his remaining energy, and he was removed as head of the monastery. Mendel had become overweight and his health had suffered during these twilight years.

8) Why is his work still important today?

It seems that on a daily basis, news reports announce important new genetic discoveries and advances.

In short, Mendel's work started a scientific revolution of such epic proportions that the life of everyone on Earth shall never be quite the same. I am convinced that Mendel's work is very likely even more important today than ever before.

Many distinctive lines of scientific study emerged from Mendel's work. First, the field of genetics, *per se*, demonstrated for the first time how genes which confer phenotypic traits are inherited and passed on to subsequent generations. Genetic inheritance is a never-ending process that maintains the continuity of all life.

Out of the curiosity held by certain early investigators about what constitutes life, there had arisen an intense interest in the nature of the genes, which then caused an even more intense interest the physical and chemical nature of the material that make up the genes, namely, the chemistry and the structural nature of DNA.

Although Mendel was to never know it, his so-called unit characters, the genes, were harbored within DNA, the great molecule of life. With the elucidation of the structure of DNA in the early 1950s, the fields of molecular biology and biotechnology exploded in the 1960s and 1970s.

The study of the complete nucleotide compositions of entire genomes, e.g., genomics, became possible during the various genome projects that materialized in the 1970s and 1980s. It has even become possible to determine the total genome sequences for each and every individual living being. Construction of a genome map and DNA sequence for each person may be relatively easy to acquire and analyze.

Moreover, it is possible in modern times to compare DNA sequences of entire genomes with those of potentially all other genomes. Interestingly, mathematics, a discipline that was near and dear to Mendel, plays an important role in genomics and other genetics-based technologies.

As a direct consequence of the studies of genetics and genomes arose the possibilities of being able to fundamentally understand at the cellular and molecular levels the basic processes of life. We now have a clearer picture of how genes, Mendel's unit characters, direct the biological activities inherent in every single living cell.

We now have a basic and rudimentary understanding of how living organisms evolve with time. Investigators have developed an appreciation of the inner workings that occur as individual living biological cells manage to produce increasingly complex living tissues, organs, and organisms.

Importantly, the field of genetics has made it possible to know which genes may be altered in a defective manner such that the normal functioning is reduced or even lost, causing genetic diseases, many of which are severe in their effects on suffering patients.

Lastly, knowledge of genetics made possible by Mendel's powerful scientific breakthrough led ultimately to the advancement in various technologies. In agriculture, for instance, improvements in molecular, gene, and animal cloning has led to better quality plants and animals with less predispositions to disease and to better yields of food.

Diagnosis of many types of diseases has become possible, such as infectious diseases, as well as in genetic diseases. In the early 1990s, the biotechnological invention called gene therapy was put to use, in an attempt to correct defective genes in genetic disease patients.

Very recently, modern investigators were able to use a new gene editing invention, called CRISPR—Cas9 (the acronym denotes a clustered regularly interspaced short palindromic repeats—caspase 9 enzyme), in order to correct a genetic disease gene for a severe heart condition in the cells of an actual human embryo.

Thus, improvements in modern healthcare therapies have become possible. Thus, genetics will no doubt continue to make new advances in our basic knowledge of life and of new technologies, many of which will have the potential to improve the quality of life for all living beings well into the foreseeable future.

Genetics has become an extremely large discipline of study. For instance, college and university courses in genetics are often required courses for students majoring in biology, biochemistry, molecular biology, bioinformatics, cell biology, wildlife biology, to name only a few. Furthermore, students can major in genetics. In fact, genetics may be represented as full academic departments housed within higher learning institutions. Entire institutes devoted to the study of genetics have been established worldwide. It was Gregor Mendel who made all of these possible.

9) What have I neglected to ask?

In his studies, Mendel had made the assumption that one gene determined one particular trait, an impression supported by the so-called "one gene, one protein" perception propounded by George Beadle and Edward Tatum, who shared a Nobel Prize in 1958 for their work regarding the genetics of metabolism. Later, the notion emerged that one gene may specify more than a single trait, i.e., one gene specifies two or more traits, a notion called pleiotropy.

Along these lines, several genes may work together to specify a range of variations for a particular trait, like, for instance, height, or color, etc. Such instances are referred to as polygenic determinism of inheritance. Multiple genes will be involved in dictating one particular phenotype.

Lastly, once the field of Mendelian genetics was firmly established in scientific circles, the question arose amongst population geneticists about how genes with dominant traits didn't just simply take over, leaving the recessive traits to eventually disappear with time. The answer came about from the population studies of Wilhelm Weinberg and Godfrey Hardy.

They proposed the so-called Hardy-Weinberg principle to account for the persistence of this genetic variation. The principle maintained that in a rather large population of individuals with no selection pressures imposed upon it, and in which mating was randomly occurring, the allelic frequencies do not significantly change, thus, maintaining the persistence of the recessive traits during the course of evolution and in the passing along of the genetic material to later generations.

Chapter 8

LOUIS PASTEUR

1) Professor Varela — I would say that almost every day, growing up I heard something about milk being pasteurized. Let's start at the beginning. Who was Louis Pasteur, and where was he born and how did he first get involved in science and experimentation?

If I may once again be so bold as to make another proclamation, it is this: I am firmly convinced that Prof. Louis Pasteur is the greatest scientific investigator who ever has lived. His scientific discoveries and contributions are legion and legendary. His experimental works have had direct and profound influences on the daily lives of virtually everyone who has ever lived since his discoveries became widely known in the mid-19th century. Pasteur's universal influence continues to this day in all circles surrounding our lives.

Louis Pasteur was born on the 27th day of December in the year 1822 in the town of Dole, in France, to parents Jean-Joseph Pasteur (his father) and mother Jeanne-Etiennette Roqui. Pasteur's interests in the sciences, research, and discovery began in 1844 while in graduate school as an MS student at the prestigious École Normale Supérieure; while both physics and chemistry courses held his interest, it is in the laboratory setting where Pasteur became fascinated with experiments and was inspired to pursue a career conducting scientific research. In 1847, Pasteur earned his doctorate in the fields of chemistry and physics from the Normale.

2) Now, why and how did he research milk, and what problem was he trying to solve?

Prof. Pasteur came to study milk in an indirect fashion by first studying wine. The story has been told over the generations since Pasteur's time that, in 1863, Parisian business people came to Pasteur, who was already quite famous by this time, requesting that he investigate their problem with soured wine, which greatly compromised its fine quality.

Pasteur's studies into the sour wine problem led him to propose to the wine makers that they practice 'pasteurizing' the wine during the winemaking process.

From this wine-saving outcome, other foods and drink were naturally considered. Prior to milk pasteurization, the problem of preserving food and drink had long been studied, first with sun-drying foods, smoking, spice additions, salting, pickling, fermentation, and heat sterilizing. Each of these procedures had their advantages and disadvantages. First, they more or less helped to curb food spoilage, with heat sterilizing, developed by Nicolas Appert, being the most efficient method. The problem with the heat sterilization method, however, was that the flavor quality may be severely affected. Clearly, the strong heat method was not good for certain foods, which may be destroyed beyond repair to be consumed.

Pasteur's method solved this food-destroying problem by focusing mainly on heating the wine and milk just slightly enough to kill microorganisms that spoiled the liquids without destroying the tastes or qualities and without having to kill all of the microbes in order to do so.

To some critics even the lowered temperatures involved in pasteurization were still too high for them, as they argued that the drinks were too much affected to be desirable for consumption.

Pasteur's main contribution to the study of milk, however, was his discovery that the microbes were the prime cause of food and drink spoilage. His pasteurization method was aimed at controlling the contaminating microbes in food and drink.

3) I know that milk basically comes from cows- is it okay to drink it from animals? Or can only animals such as goats and sheep drink parental milk? And how and when did humans start to drink "pasteurized" milk?

Milk is a nutritious substance, filled with lots of ingredients which are thought to be good for the general health of mammals, including humans. Indeed, milk has many good health benefits, such as, for instance, providing protective maternally derived antibodies to infants whose immune systems may not yet be fully developed. Milk is a good source of protein and good quality carbohydrates, such as lactose, the so-called milk sugar that confers a desirable sweet taste to milk and other milk-based foods.

There has been a long-standing controversy whether one should consume raw milk directly from animals (i.e., unpasteurized milk). Others argue that one should first pasteurize raw milk prior to its consumption. One the one hand, proponents of raw milk drinking insist that its taste, nutritious quality and safety are all maintained. On the other hand, proponents of pasteurized milk drinking insist that contamination with infectious disease causing microorganisms is all but eliminated, ensuring its safety from both morbidity and mortality. The controversy seems to be rooted in the issues associated with putative costs in terms of adhering to regulations, management, reporting, and in the pasteurization process.

On historical and scientific bases, the consumption of raw unpasteurized milk that might be contaminated with microbial pathogens has been known to be directly responsible for causing infectious disease. For instance, cases of tuberculosis, anthrax, cholera, scarlet fever, staphylococcus, typhoid fever and diphtheria have all been definitively documented as having been transmitted by contaminated raw milk.

More recent work, conducted in our research laboratory, showed that when wild-type *Escherichia coli* bacteria are cultured in pasteurized milk, susceptibility to antimicrobial agents is reduced for a variety of drugs. Molecular genetic analysis of this effect showed that the reduced drug susceptibility involved a genetic element on the *E. coli* chromosome, called the *mar* locus, an operon that, when turned on, shows enhanced antimicrobial resistance to multiple anti-bacterial agents. It is known that even pathogenic strains of the *E. coli* harbor the *mar* locus, as well.

Historically the consumption of pasteurized milk began soon after Pasteur developed his pasteurization methods, especially after the publication of a refined procedure in 1866 by Franz von Soxhlet, a German chemist. He knew of Pasteur's work with wine pasteurization and proposed that milk and other drinks be pasteurized, too. Today, not only are beverages like milk, juices, and alcoholic drinks routinely pasteurized, but so too are many other food products, such as yogurt, cheeses, ice creams, pickles, etc.

4) Now what exactly do we mean by pasteurized milk? And how did Louis Pasteur come upon this process?

Pasteur's work with milk pasteurization arose out of his earlier studies with fermentation in the mid-1850s and with wine in the early 1860s. Pasteur first studied milk in 1857 when he found that milk sours by turning the milk sugar lactose into lactic acid, in a process we now know as fermentation.

At the time, however, Pasteur had correctly claimed that the milk souring process was performed by the microbes, which he called ferments, contained within the milk. He was widely criticized, however, for claiming that microbes could do chemistry, a notion now known as biochemistry—living beings partaking in conducting chemical reactions. Most of Pasteur's critics held that the milk-souring process was due solely to nonlife-based chemical reactions, independent of the living microbes. We now know that the milk souring process involves microbial metabolism in which the lactose sugar is broken down into glucose and galactose by microbial enzymes, broken down further by oxidation to pyruvate molecules, and then fermented into acidic end-products, like lactic acid, by other microbial fermentation enzymes. Pasteur had been correct all along.

Pasteur had been the first to postulate that microbes conducted biochemistry by souring milk and wine and, importantly, by actually making wine and other alcoholic beverages like spirits, beers, liquors, etc. Out of these newly formed biochemical ideas, Pasteur made the greater conceptual leap that if microbes could make milk and wine sick (spoilage by becoming soured), perhaps microbes could also make people sick. Hence, Pasteur formulated his amazing historically important hypothesis that microbes can cause infectious disease. The hypothesis became known as the so-called germ theory of disease.

The process of pasteurizing milk is meant to prevent its spoilage. It is not necessarily a sterilization process but rather it is a pathogen-killing one, with many millions of microbes remaining alive. Pasteurization itself involves heating the raw beverage to temperatures that kill microbes which, under normal circumstances grow best at temperatures (37°C, 98.6°F)—the same optimal temperature of most mammals, including humans; such moderate-temperature loving microbes are usually the most efficient pathogens towards humans and other animals. In the past, the pasteurization method involved the batch heating of large quantities of milk to 63°C for about 30 minutes.

In modern times, the so-called flash pasteurization method is practiced. This process heats the milk to 72°C as it is passed through thin tubes, but for only a very short time period, about 15 seconds. Another method, called ultra-high-temperature pasteurization heats the milk to temperatures as high as 135°C, but for only about one second! This latter pasteurization process is controversial as critics claim the taste of milk is adversely affected by this high-heat method.

A variation of the ultra-high-temperature pasteurization method is the ultra-high-temperature sterilization technique in which all of the microbes in the food or drink are killed by using superheated steam at about 140°C for 1 or 2 seconds. For restaurant connoisseurs who use the little packages containing dairy creamers for their hot beverages, the creamer substances have been ultra-high-temperature sterilized, rather than pasteurized. These sterilized foods and drink have a relatively longer shelf life.

5) I think he also studied fermentation (which I believe has something to do with grapes) can you clarify?

Pasteur's first experience with grapes involved also his first research studies, in 1844, as a young scientific investigator. He studied a compound found in grapes, called tartaric acid. His first major discovery showed, optically speaking, that the tartaric acid crystals formed in mirror-image isomers of each other and rotated the plane of light; and these tartaric acid crystal variant mixtures, he called racemic, which was derived from the Latin term *racemus,* for grape.

Pasteur studied fermentation from two standpoints. The first being lactic acid fermentation and the second being alcoholic fermentation.

Pasteur's interest in lactic acid fermentation in 1857 came about due to its convenience with which to study. Lactic acid was produced in mass quantities from beet juice and as a by-product of alcoholic fermentation. In his experiments, Pasteur showed that a microorganism, nowadays thought to be a bacterium called *Streptococcus lactis*, when grown in a culture medium containing chalk, which was an alkaline substance, plus sugar and an extract of brewer's yeast for food, produced lactic acid in large quantities, after an incubation period. In essence, Pasteur was claiming that a living being was undergoing fermentation; that is, a microbe was doing biochemistry by making lactic acid. The publishing of this work in 1857 has long been considered by historians and scientists alike to be the so-called birth certificate for the establishments of the scientific fields of microbiology and of biochemistry.

Pasteur's work with alcoholic fermentation showed that yeasts (then called ferments) were actually living organisms and that they themselves could produce the ethanol inherent in the process of alcoholic fermentation. The traditional process for alcoholic fermentation has historically involved the grapes that you spoke of above, which was a good source of the sugar needed to conduct the process. When Pasteur used grape juice as a source of sugar, he noticed bubbling, indicating that carbon dioxide was being produced.

Furthermore, Pasteur showed experimentally that when his yeast ferments had water, sugar, ammonium salts and trace elements, the yeasts could be coaxed to not only produce alcohol efficiently, but that the yeasts, too, were growing by taking up various elements from the sugar sources.

Pasteur reasoned that the yeast cells may have been producing the wine all along; this led to his formulation of the hypothesis that microbes were conducting chemistry.

Pasteur's studies of fermentation led also to a phenomenon which biochemists nowadays still appreciate. The phenomenon is known as the 'Pasteur effect,' a term coined by Prof. Otto Warburg, another famous biochemist, in which Warburg meant it to be the inhibition of the glycolytic pathway when in the presence of oxygen. In 1861, Pasteur made the astute observation that oxygen was inhibitory to fermentation. Later, in 1875, he boldly proclaimed that fermentation was life without the necessity for oxygen. Life could occur without oxygen as far as fermentation was concerned. The Pasteur effect is still included in modern biochemistry textbooks.

6) Now, what is the "germ theory of disease" and how is he linked with it?

One of Pasteur's most famous hypotheses was the idea that microbes caused infectious disease, the so-called germ theory of disease. Interestingly, this theory had its origins with his work on alcoholic fermentation, which, in turn, had its roots when local winery owners approached Pasteur concerning the problem of their vintage wines having gone off, being soured and becoming more like vinegar than wine. He studied the healthy versus the sick wines with the microscope; in the good wine, he found large yeast cells and in the unhealthy wine, he noticed the yeast cells plus smaller bacterial cells.

Pasteur reasoned that the bacteria were spoiling the wine; this led to his formulation of the famous germ theory of disease. His reasoning was that if the microbes made wine sick, then perhaps microbes were making people sick, too. Unfortunately, Pasteur did not do any experimental work to provide any evidence for this germ theory of disease.

The first evidence provided in support of Pasteur's germ theory of disease was published in the 1870s with Dr. Robert Koch, who studied anthrax disease. Koch developed a methodological approach still in use today, called Koch's Postulates, in which he showed that a rod-shaped bacterium, now called *Bacillus anthracis*, was the causative agent of the anthrax disease, thus definitively demonstrating Pasteur's theory that microbes could cause disease. Koch was later to demonstrate in 1882 that the bacterium called *Mycobacterium tuberculosis*, was the causative agent of the dreaded consumption, a discovery for which Koch was to receive the Nobel Prize in 1905.

7) I remember from college the terms "spontaneous generation" which was later disproved- germs, bacteria and viruses - DO NOT just appear out of thin air - How was Louis Pasteur involved in this?

The historical term spontaneous generation has its origins reaching as far back as 343 B.C. with Aristotle, who first proposed the idea in his writings on the topic of the natural sciences. Aristotle wrote that living organisms could arise spontaneously out of non-living materials, like living flies arising spontaneously on non-living rotting meat. In short, the idea was that life arose out of non-life. The modern term for this phenomenon is abiosis—life from non-life. The abiotic notion of spontaneous generation remained in academic circles for almost two millennia. No one since Aristotle had recorded an experimental examination of spontaneous generation—that is, no one had actually bothered to put the idea to the test—until 1668, when Francesco Redi had published the first experimental analysis of the phenomenon. Redi's published evidence argued against the idea of spontaneous generation, at least for flies on meat, that is. In 1750, John Needham had tested the spontaneous generation in the then newly discovered microorganisms (bacteria) on mutton gravy where he failed to adequately heat the test medium to completion to eliminate contaminating heat-resistant endospores. Upon cooling, the bacterial endospores germinated and grew to visibility. Thus, Needham erroneously argued that spontaneous generation occurred in bacteria. This was essentially the state of scientific affairs when Pasteur entered into the arena of spontaneous generation in the early 1860s.

Pasteur's entry into the detailed study of spontaneous generation started with his curiosity about his studies on milk spoilage into lactic acid, fermentation of sugar into wine alcohol by yeast, and wine spoilage into vinegar. He had established that microbes were responsible for each of these fermentative processes. Next, he contemplated the origin of the microbes involved in these processes. He had wondered where the microbes came from.

Although Pasteur addressed spontaneous generation by conducting many experiments, two classic ones deserve mention. The first involved grapes and yeast. He obtained grapes that were free of yeast and showed that they could not ferment, while grapes injected purposefully with the yeast freely underwent fermentation. This work, thus, demonstrated that the yeasts were responsible for conducting the biochemical process of converting the sugar within grapes into alcohol. A comment of Pasteur's work by the famous physiologist Claude Bernard was made in which he (Bernard) speculated that yeasts underwent spontaneous generation. This led Pasteur to conduct in the early 1860s what is considered by many to be the most famous scientific experiment ever performed—the so-called swan neck flask experiment.

Pasteur put the spontaneous generation theory directly to the test with microorganisms and the swan neck flask. Pasteur was said to have actually constructed these glass vessels himself. The swan neck flask has a round bottom with an elongated neck that winds around like the neck of a swan—hence, the name, swan neck flask. The neck of the flask allowed air to move about freely as air was considered the vital driving force for the spontaneous generation process to occur.

A simplified explanation of the famous swan-neck flask experiment follows here. First, Pasteur filled his swan neck flasks with liquid broth culture media, a so-called yeast infusion broth. Next, Pasteur heated by boiling the broth-filled swan necked flasks, in order to kill off any microbes contained within the broth—luckily for Pasteur, in the broth there were apparently no confounding heat-resistant bacterial endospores that would have survived the heating. Non-boiled flasks were included as controls. Next, he allowed the sterile swan neck flasks to incubate, allowing both air and time for any putative microbes to generate spontaneously in the broth and grow.

After sufficient time, about 2 or 3 days, there was no growth in the boiled flasks while the unheated flasks had obviously visible microbial growth. Next, he actually broke the necks of the heated flasks without microbes and allowed the microbes from the air to fall into the broken flasks to encounter the sterile culture broth and grow after 1 or 2 days. Pasteur published these initial findings in a scientific journal on the 6th of February in 1860.

With time, he followed up on these preliminary findings with additional and more-sophisticated experiments, using other culture media like milk and urine and invoking variations in the swan-neck flask technique, but the damage to spontaneous generation had been done. With only a few stubborn holdouts remaining, spontaneous generation was, nonetheless, no more. Spontaneous generation was, as you stated above, disproved finally.

8) I also understand that Pasteur was somehow involved with the rabies vaccine. I understand that being bitten by a rabid dog can result in death. What was his involvement in this realm?

Indeed, throughout most of written history, rabies has been a terribly dreadful disease. Death by rabies is frightening and slow. It is extremely rare to survive rabies, even after a vigorous course of treatment. Only in recent times, has there been a handful of documented cases in which patients survive rabies after infection, as seen, for example, with chemical induction of a coma.

Treatment of rabies in Pasteur's time was equally frightening. Pasteur had recalled a childhood memory, from 1831, of the terrifying course of treatment for unlucky victims—namely, the cauterization of the flesh wound with a red-hot iron where the rabid animal had bitten the patient. The treatment frequently failed, and the victim often died anyway, usually after having first gone insane.

The recurrence of the dreaded rabies in the region where the grown up Pasteur was living, in 1885, and his success with having developed an anthrax vaccine, compelled him to take up the study of rabies for the ultimate purpose of inventing a rabies vaccine.

First, Pasteur began a series of experiments meant to cultivate the rabies agent; and after a long series of setbacks, he found that rabbit spinal cord was a good medium with which to maintain the rabies microbe. One of the most famous oil paintings of Pasteur is the portrait of him sitting next to his laboratory bench leaning on a book and holding a glass vessel containing a hanging piece of rabbit spinal cord laden with the rabies agent—his is shown looking intensely at his rabies cord. Using this spinal cord method of cultivation, Pasteur was able to isolate and propagate the most virulent deadly form of the rabies agent, testing his virulent rabies on dogs in the laboratory, giving them rabies rather quickly.

Next, Pasteur used this highly virulent rabies strain to begin the process of attenuation—a technique of drying the spinal cords and weakening the rabies microbe in order to lose its virulence but at the same time retain its immunogenic properties, the latter of which were needed to ensure a good, i.e., protective, vaccination. He initially began with the least virulent forms of the rabies on the spinal cords and used increasingly virulent rabies forms to inoculate the laboratory dogs.

After inoculating the dogs with the attenuated rabies vaccine, he was gearing up to test its efficacy by then injecting the dogs with the original highly virulent rabies agent to see if the vaccine protected the dogs. However, a history-changing event occurred before he could do so.

A little boy named Joseph Meister, a nine-year old at the time, had been bitten and severely injured by a rabid dog, and the prognosis was not good. His mother had known of Pasteur's rabies work and had made a convincing argument to Pasteur himself to give young Joseph his yet untested rabies vaccine. Only after great trepidation and reluctance, Pasteur agreed to provide the requested rabies vaccine to young Joseph.

On the 6th of July, 1885, the treatment had begun of the Meister boy, receiving the same course of inoculations as the dogs had been given, starting with the least to the most virulent but attenuated forms. The method has since been called the 'Meister method' of treatment. The wait-and-see period afterwards was an agonizing one for Pasteur and co-workers, namely Dr. Emile Roux, one of his colleagues who had resigned as a result of having witnessed the medication of an untested vaccine on a child. He later returned to the lab when word of the successes with the vaccine became widely known—Pasteur was never to know how Roux had felt about the incident.

Young Mr. Meister lived to survive the entire ordeal. In fact, he lived to be an elderly gentleman. Throughout the years after the rabies vaccine incident, the Pasteur family had kept touch with the Meister family. As a matter of fact, Pasteur had provided funding for young Meister's education and even made arrangements to provide employment as gate guard at the Institute of Pasteur in Paris, France. Legend has it that as a guard, Meister had refused admission to German Nazi visitors who wanted to visit the tombs of Pasteur and his wife, Marie Laurent Pasteur, days after the German invasion during World War II. Sadly, shortly after the invasion, Mr. Meister committed suicide at home, apparently because of the guilt he felt for having sent his family away during the war, inaccurately believing they had perished because of him.

9) During his lifetime, Pasteur must have kept very busy- as I hear he also studied Pébrine disease. What exactly is Pébrine disease and what are some of the signs and symptoms and what was Pasteur's involvement in this?

Pébrine is a devastating disease of silkworms in which they are infected with a parasitic protozoan. Prior to Pasteur's involvement with the Pébrine, the disease was a virtual mystery. There were no obvious signs and symptoms—only death of the silkworms. What was known about Pébrine was that it killed the silkworms and, thus, an outbreak of Pébrine nearly destroyed the entire silk industry in France in 1865. When people from the silk industry begged Pasteur to study it, he protested that he didn't know anything about silkworms, and certainly no one knew what caused the Pébrine. Nevertheless, industry representatives and close friends alike insisted he was just person right for the job of figuring out the cause of the malady—he was a scientist with no preconceived notions and, therefore, he wouldn't be biased towards or against any misleading paths.

He carefully studied the anatomy and physiology of the silkworms, comparing the healthy versus the diseased worms. It was slow and painstaking work—it took him three years, but he eventually identified two causes, one pertaining to the discovery of the infecting protozoan in the eggs of diseased silkworms, and the other, a pathological condition named flacherie, having to do with poor quality nutrients and the starving silkworm. Pasteur even provided various methods that those in the silk industry could employ in order to eradicate the two maladies. Pasteur is credited with having saved the entire silk industry and preventing the spread of the Pébrine to the rest of Europe.

10) What have I neglected to ask?

Interestingly, historians of science, a prominent one being Dr. René Dubos, who wrote an elegant little biography of Pasteur, frequently state that Pasteur had come very close to discovering antibiotics. During his studies of the bacterium known to cause anthrax, *Bacillus anthracis*, Pasteur had observed that when the *B. anthracis* cultures had been mixed with contaminating soil microbes, probably with bacterial species of the genus *Streptomyces*, the anthrax bacteria grew less well. The work predicted that some growth-inhibitory substance produced by the soil bacteria was responsible for the growth effects on *B. anthracis*. Had Pasteur gone one step further, he may have been the first to discover an antibiotic, instead of Alexander Fleming with his discovery of penicillin in 1928.

Pasteur later was to be a strong proponent of Listerism, which was the use of antiseptic agents during surgery and in the dressings of surgical wounds.

In another less well-known experiment conducted by Pasteur and briefly mentioned in his published 1857 paper on lactic acid fermentation, he had reported on the inhibitory actions of onion and garlic juice on the growth of his famous ferments, which we now know to be yeast microbes. In some of our recent work conducted in our laboratory, we found in the published literature that onion and garlic share a similar bioactive principle called allyl sulfide. Our results showed that the garlic extract and allyl sulfide inhibits a drug efflux pump, called EmrD-3, from *Vibrio cholerae*, and that garlic extract made other antibiotics work better in a process called synergy. In our published paper, we were delighted to mention Pasteur's inspiring work.

On the 19th day of October in 1868, Pasteur suffered a massive stroke. At first, he had been severely debilitated, but a few months later, he recovered sufficiently enough to begin walking again. Nevertheless, Pasteur was left with a long-lasting partial paralysis on one side of his body. Again, on the 23rd of October in 1887 Pasteur suffered another stroke. This time he was unable to speak and had to have his son read a scheduled speech for him while Pasteur remained silent. He did manage to recover, but he did suffer from partial paralysis for the remainder of his life. He continued to work hard despite his condition, as he had done so all of his life. On the 25th of September in 1895 he passed away, leaving behind an overabundance of amazing discoveries, any one of which in their own right could easily have made him a well-respected and famous scientist.

Chapter 9

JOSEPH LISTER

1) Professor Varela, please excuse my poor attempt at humor above, but I would like to ask some questions about Joseph Lister, who lived in the 1890's and who apparently was known for the use of antiseptic during surgery. What do we know about this Joseph Lister and his early life?

I am positive that you are not the first one to have made that particular humorous connection between Dr. Joseph Lister and the Listerine mouthwash. Although Lister himself did not develop the chemical agents that constitute Listerine, the mouthwash antiseptic was nonetheless named in his honor by others in 1879.

Lister was the first surgeon known to employ the use of antiseptic chemical agents during surgeries. The honors in his name, however, did not stop with the Listerine. In 1927, the scientific name of a pathogenic bacterium, *Bacterium monocytogenes,* was renamed as *Listerella hepatolytica* and then later changed to *Listeria monocytogenes*, after the then very famous Joseph Lister.

The disease caused by this dangerous microbe is referred to as listeriosis. Likewise, a slime mould was named after Lister, called *Listerella paradoxa*, and was classified under the scientific family name of Listerelliidae. The technique of using antiseptics during surgery, a method that you referred to in your inquiry, is known as Listerism. Numerous buildings, a hospital, and even an entire institute have been named after him.

The honors did not end with Lister's name. Lister has been featured in countless biographies, science books, popular films, documentaries, stamps, monuments, and memorial funds.

Joseph Lister was born in a mansion called Upton House on the fifth of April, 1827, in West Ham, a countryside town that resided in a borough of London called Newham, in England. The members of the prosperous Lister family were Quakers, known at the time as the Society of Friends. Lister's mother, Isabella Harris, has been described by biographers as kind and loving. An early influence on the young Lister was his father, Joseph Jackson Lister, a wine businessman and amateur microscopist who was also a Fellow of the Royal Society and who taught the kid Lister how to use the microscope. In addition to English, the young Lister spoke German and French.

2) Since he seemed to be involved with surgery, I assume or presume that he had some kind of medical training. Where did he go to school and did he have a specialty of study?

While attending Grove House, a Quaker school in a town north of London, called Tottenham, Lister took an interest in anatomy and consequently left the school to begin pursuit in the study of medicine to become a surgeon. As Cambridge and Oxford did not accept Quakers, Lister first went to University College in London, where he struggled with a case of smallpox and then suffered a mental breakdown. His breakdown was attributed to a premature reentry to school before being able to make a full recovery from the dreaded smallpox. He did eventually graduate in 1847 with his BA degree in botany. Lister then returned to University College and studied medicine. Here, Lister flourished, taking his bachelor's degree of medicine (MB) in 1852 while earning academic honors. Next, Lister was admitted to the Royal College of Surgeons, England and later became an assistant to the then famous James Syme, Prof. of surgery, housed in the Royal Infirmary at the University of Edinburgh, Scotland. The assistant and professor became lifelong friends. In fact, Lister courted and later married Syme's daughter, Agnes. In 1860, the Listers moved to Glasgow, where Lister became Regius chair of surgery. They then returned to Edinburgh in 1869 where Lister became professor of surgery. In 1877, Lister became chair of surgery in King's College Hospital housed in London. He retired from this institution in 1882 and died at the age of 84 on February 10, 1912.

3) Now, the key question seems to be- what led him to believe that the use of antiseptics during surgery would save lives?

Prior to the use of antiseptics during surgeries in 1865 by Lister, the general state of surgery was fraught with staggering mortality rates, but not after first succumbing to terrifying diseases beforehand. Anesthetics, like ether, developed in 1846, had made surgery a painless and, thus, a significantly less rapidly performed medical procedure. Prior to the use of anesthesia, surgical procedures needed to be as quick as humanely possible, in order to minimize the pain to the patient. With anesthesia, surgeons could now take their time, carefully probing deeper into living tissues of the patient. Sterile surgical technique had not yet been implemented. Although surgery had now became painless with the anesthesia, more often than not, the patients still succumbed to post-operative infections.

In fact, prior to Lister, post-op infection was actually a completely expected part of the normal consequence of having undergone surgery; a post-op infection was routine. The best case scenario might be a relatively mild, limited infection that would need only pus drainage (referred to as laudable pus), leaving, perhaps, a grotesque scar. The so-called laudable pus was considered a positive sign, nonetheless, that the post-surgical patient might actually live.

The most common scenario, however, was certainly not the best one as described above with the laudable pus. On the contrary, it was death, but often not before first acquiring a more severe post-op infectious disease, like the erysipelas (severe skin rash), the pyremia (high fever—a very bad sign), the putrid gas gangrene (a foul-smelling condition indicating an amputation was soon to be at hand), and the dreaded and putrid sepsis (a blood infection, the same kind that had confounded Semmelweis and countless mothers giving birth).

Treatments of post-op infections were often frightening and, unfortunately, not effective. They included amputations of limbs with dead tissue, pouring boiling oil onto open wounds, and cauterization of infected wounds with red-hot irons. The patient would still acquire the infection and then die. This was essentially the medical state of affairs for surgeons during Lister's time, and he agonized about it for years.

Surgeons of the time had their own idea to explain the post-operative infection and its deadly state of affairs. Because of the miasma (foul-smelling air that was thought to cause disease), the oxygen in the air was believed to play a central role. Open wounds, like those made in surgery or those made by violent trauma, were obviously exposed to the air. If such was the case, then post-op infections and the deaths associated with them, were simply considered to be unavoidable outcomes.

Lister had a few nagging reservations, however, about the Miasma oxygen exposure theory of post-surgical infections. First, Lister wondered why oxygen that's presumably embedded deep within tissues didn't always become infected. According to the oxygen theory, it should have also been infected; it was not. Second, Lister wondered why lungs that had been perforated by broken ribs didn't always cause infection in surrounding tissues. According to the oxygen theory, exposure of nearby internal tissue to lung oxygen should cause spontaneous infections there, too. It didn't.

Third, Lister made the astute observation that for bone fracture injuries in which the skin was broken and bone was exposed, infections arose; while those bone fracture injuries in which skin was intact and bone was not exposed, i.e., so-called closed bone fractures, no infections arose. Thus, Lister reasoned that open wounds, with exposed bone fractures, may have experienced the entry of some foreign agent, other than the oxygen. Next, Lister tied together the foreign agent wound entry idea with another idea formulated by the great Prof. Louis Pasteur, which held that microbes called bacteria were the germs that caused disease of wine during wine fermentation. Lister reasoned that bacteria were the foreign agents entering the open wounds and causing infections and death.

What Lister did next to the microbes in the surgical wounds was to result in the changing of medical history forever. He set out to kill these wound entering microbes.

Lister sampled infected wounds and using the microscope, he demonstrated that they were indeed teeming with bacteria. Understanding that heat was not practical for treating the infected wounds, he tried a series of chemicals applied directly to the surgical openings. None reduced the infections. He then tried a chemical called carbolic acid. Lister's choice of carbolic acid came about from reading a report about how its use in the regional sewer system reduced the foul smell of putrefaction.

First, Lister observed that in culture the carbolic acid killed the wound bacteria. Next, although the details are sketchy, in March of 1865, the very first application of the carbolic acid to a patient did not go well, and the patient apparently died. The first successful carbolic acid application, however, did not occur until August 12, 1865. Lister prepared wound dressings that were made of gauze, soaked them with the carbolic acid, applied them directly to a freshly bone-fractured and open leg wound of another patient, and covered the antiseptic-laden dressed wound with tin foil, to prevent the antiseptic from evaporating. The patient was 11-year old James Greenlee, who had been run over by a horse-drawn wagon just hours before Lister's new history-making treatment was applied. Lister changed the carbolic acid soaked gauze dressings periodically. Amazingly, not only did the young Mr. Greenlee fail to have a wound infection and not need a leg amputation, he also failed to die. After a few weeks he was released from the hospital in a perfectly healthy state. Feeling invigorated, Lister similarly treated 10 additional patients with his life-saving antiseptic, and 9 of these patients lived, demonstrating that the exposed bone fractures could be treated with antiseptics.

Lister focused his efforts next on patients who had abscesses in their wounds. He prepared a putty containing carbolic acid, among other materials, and he applied the carbolic acid-putty mixture to the abscessed wounds and was met with success. The abscessed wounds healed. Lister published these findings in 1867.

Encouraged by these successes, Lister then used his antiseptic on actual surgical patients. He poured the carbolic acid mixed in linseed oil into surgical openings. He even took to handwashing with the carbolic acid prior to surgery; he soaked the surgical instruments with the same antiseptic. He even sprayed his carbolic acid throughout the whole of the surgical theatre. Lastly, he dressed the post-operative incisions with gauze containing the carbolic acid. The incidence of post-operative infections greatly diminished.

Lister published these latter findings in a series of journal articles between 1867 and 1875. These publications, recorded in history, outline the very first successful application of antiseptics to the practice of surgery.

4) What was the initial reaction to this in the world of medicine?

Unfortunately, the initial reaction to Lister's surgical antisepsis technique was plagued with its immediate rejection by the English surgeons outside of his home institution at Glasgow. There are several postulated reasons for this rejection; and chief among them was the fact that the carbolic acid was actually caustic and irritated the skin, including that of the surgeons' hands. Additionally, although the patient had greatly diminished chances of acquiring post-op infections, the surgical incisions, nonetheless, had to heal from the tissue damage caused by the carbolic acid burns.

Furthermore, the mixture of the carbolic acid and the linseed oil prevented a clear field of vision of the surgical openings during surgeries. This made it difficult for the surgeons to see clearly during surgery. Third, the spraying of the carbolic acid throughout the operating room resulted in its pungent and disagreeable odor throughout the surgical facilities. In addition, due to its volatile nature, the carbolic acid tended to evaporate, requiring more frequent post-operative changes in bandage dressings than before. This, on top of an already complicated and cumbersome procedure for bandage changing, became a rather bothersome practice.

5) How long did it take before this was widely accepted?

Unlike the case with Ignaz Semmelweis, who never in his lifetime received acceptance for his handwashing method, Lister was exceedingly fortunate in that acceptance of his antiseptic technique came sooner rather than later. Lister enjoyed numerous honors, prizes, tributes and dedications. He was well-respected and greatly admired. Even today, periodic tributes are routinely published in scientific circles. Nevertheless, acceptance of Listerism was not immediate.

In 1869, the Listers moved to Edinburgh, where he had to convince his colleagues there of the benefits of his surgical antiseptic technique. Interestingly, shortly after his arrival, Lister had been summoned to the Balmoral Castle to treat Queen Victoria, who had an axillar abscess. He applied the antiseptic technique during her surgery, and she survived, saving the Queen. Years later, in 1883, she knighted him.

The first inklings of acceptance for the so-called Listerism began with the French and Germans in 1870 during their war with each other; both sides uniformly adopted Lister's antiseptic practices, in order to reduce battlefield mortalities from infections. After the war, the military surgeons took the successful antiseptic techniques to their homelands and successfully applied them to their civilian patients, as well. The practice quickly spread throughout most of Europe.

Acceptance of Listerism in the U.S. came more slowly, even after Lister, though widely celebrated at the time with honors and accolades throughout Europe, toured the U.S. from coast-to-coast spreading his message. In the U.S., Lister was met with skepticism. Medical education in the U.S. largely ignored the newly emerging basic medical science disciplines. The Americans at the time did not have established practices of sterile technique or of experiment-based medical research, and prominent surgeons openly scoffed about Listerism, even in their popular medical textbooks. Some U.S. surgeons claimed to have tried the antiseptic technique, only to have it fail miserably.

It took the assassination of its president, James A. Garfield, for acceptance of Listerism to take hold in the U.S. On July 2, 1881, Garfield in Washington, D.C., was reportedly shot twice by Charles J. Guiteau, grazing his (Garfield's) shoulder with one shot, but leaving one of the bullets lodged in the president's back with the other shot. Garfield's physician, Dr. Willard Bliss, a long-time friend and non-believer of Listerism, poked around in the bullet wound with apparently dirty hands and instruments, and it is likely that this is the primary reason why Garfield's bullet wound became severely infected.

Sepsis set in quickly but death for Garfield came slowly, a full 80 days after being shot. Autopsy confirmed death by the sepsis, and the unfortunate incident sparked a renewed interest in Listerism. The final acceptance in the U.S. was complete when a German physician, Dr. William Stewart Halstead, at Johns Hopkins, implemented Lister's antiseptic technique during surgery, and it eventually became widespread throughout the country.

Finally, acceptance of Lister's method in England took place only a full 2 years after his move to Kings College, London, in 1877. Debate about Listerism was contentious during this period, and it was not until a scientific conference, called a congress, was held in 1879 at St. Thomas Hospital, where data were presented showing the improvements (i.e., reductions) in morbidity and mortality rates associated with the use of antiseptic methods during surgery.

Since this congress was first held, surgeons have widely adopted the practice of Lister's surgical antiseptic method. It further led to the adoption of the sterile technique during surgeries in order to prevent microbial wound contamination of surgical patients in the first place, a process called asepsis.

6) What other discoveries did he make or what else was he known for?

As a student of surgery, Lister published two papers, one dealing with eye tissue and the other in muscle tissue, both studies of which were based on his use of the microscope. Additionally, Lister is known in microbiology circles for developing a technique in 1879 to isolate bacteria in pure cultures based on their fermentation behaviors, although this latter approach was not as straightforward as those of Koch or even Pasteur.

It did, however, lead later microbiologists to refine Lister's isolation technique to develop the so-called 'most probable number' method to estimate the number of individual bacteria contained within a liquid sample. Lister is also known to have developed several surgical techniques, chief among them being the so-called chromic catgut method, in which the degradable catgut was soaked with chromic salts and carbolic acid and then used in surgery. The chromic catgut technique is still in use today.

7) What have I neglected to ask??

I think perhaps one might be curious to consider the nature of the microbe that was thought to be the causative agent of the sepsis in Lister's time. Sepsis is a condition in which bacteria actively grow in the blood, as opposed to bacteremia, which is merely the presence of bacteria in the blood. The dread sepsis in Lister's day was primarily attributable to the bacterium called *Staphylococcus aureus*, a Gram-positive microbe that produces a golden pigment, visible on Petri plates with culture media as bright golden colonies of bacteria.

Variants of this microbe are known as methicillin-resistant *S. aureus* (MRSA) and are considered serious pathogens today. A number of years ago, our laboratory studied a MRSA clinical isolate and cloned a gene encoding a multidrug efflux pump, which we called LmrS and which confers resistance to several antimicrobial agents. We recently published another paper in which we were able to inhibit the physiological activity of this LmrS multidrug efflux pump using a cumin spice compound called cuminaldehyde.

The bacterium actually has about a dozen of these antimicrobial efflux pumps. In fact, *S. aureus* has a series of other antimicrobial resistance systems, making treatment of infections with this microbe increasingly problematic. Clearly, much work remains to be conducted in order to ultimately control the devious nature of this historical sepsis-causing microbe.

Chapter 10

ROBERT KOCH

1) Robert Koch apparently discovered the anthrax disease cycle around 1876 and then the bacteria responsible for tuberculosis later in 1882 and then followed up with an investigation into cholera a year later in 1883. In terms of discoveries and investigations, he has to be one of the most productive scientists of all time. Your thoughts?

I completely concur with your assessment of the great Prof. Heinrich Herrmann Robert Koch. His investigative contributions to science were vast, and in their own individual circumstances, each of Koch's quite numerous discoveries helped to revolutionize the progress of the scientific research in very productive and long-lasting manners. Humankind is benefiting tremendously from Koch's scientific findings, even to this very day.

You mentioned above three very important bacteria, each of which are of tremendous importance in the infectious disease arena of medicine.

Regarding the so-called anthrax life cycle, Koch showed in 1876 how these bacteria, called *Bacillus anthracis*, form environmentally inert and hardy structures called endospores and how the endospores could revert by germinating back to actively metabolizing and growing states. Koch's anthrax life cycle studies helped to find new avenues for control of infectious disease.

Regarding Koch's tuberculosis work, for which he eventually was to earn the Nobel Prize, he demonstrated in 1882 that the tubercle bacillus, a bacterium called *Mycobacterium tuberculosis*, was the causative agent for this terribly dreadful disease—a malady that 'consumed' its human victims—known as the Great Consumption.

Regarding Koch's work on cholera, in 1884 he was able to grow the causative agent in a pure culture for the first time in history. The bacterium, called *Vibrio cholerae*, has wreaked havoc numerous times throughout human history with great waves of devastating pandemics, causing terrible losses of life during each of the cholera pandemics.

Dr. Koch was key to many additional and quite significant scientific discoveries. A few key contributions made by Koch are summarized below.

As briefly mentioned in the chapter on Louis Pasteur, Koch was the first investigator to demonstrate experimentally the validity of Pasteur's hypothesis that 'germs' cause disease. In 1876, Koch published his famous paper using *B. anthracis* to show that the bacterial 'germ' caused the deadly anthrax disease in animals.

Importantly, in 1883 Koch established the efficacy of his so-called 'postulates' to extremely good use. As such, Koch's postulates represent a definitive and widely accepted method for making an association between a given disease and a microbe as a direct cause of that disease. Clinical infectious disease specialists universally regularly practice Koch's postulates to this day.

Another important contribution still in use today is Koch's pure culture plating technique for isolating bacteria from discrete colonies of bacteria on Petri dishes containing, e.g., nutrient medium and agar. Bacterial isolates are regularly obtained today using essentially the same technique as Koch in the early 1880s.

Koch advanced many novel improvements in the field of microscopy during the 1870s and 80s. He developed many new chemical-based staining techniques to better visualize bacteria under the microscope.

Combining the then new technology of photography with the relatively older technology of microscopy in 1877, Koch was the first to publish a photograph of a magnified microbe. This methodological approach devised by Koch is still of great relevance today, both in research and in education.

During the late 1870s, Koch invented the so-called slide technique of attaching bacteria to glass microscope slides. The Koch slide technique is taught in any laboratory course dealing with microbiology to this day and in virtually the same way that Koch had invented it.

Koch was a key figure who used experimental studies, starting in 1878 and onward, to provide supporting evidence for Lister's ideas regarding wounds infected by bacteria. Thus, the idea that bacteria could infect wounds led the way for improvements in sterile surgical technique.

Along with others like Lister, Pasteur, and Semmelweis, Koch dabbled in honing the sterile technique in 1881 as applied to the microbiology research laboratory. Such sterile practices were to be widely adopted in the medical and surgical settings.

In 1890, he developed a so-called tuberculin test, which was geared to determine whether one had been exposed to the tubercle bacillus by evaluating an immune reaction against the bacterial agent, adjacent to the skin's outer epidermal layer. Although first touted as a cure for tuberculosis, the tuberculin test instead became a rather useful diagnostic tool.

Koch made suggestions in 1892 for the improvement of water quality in an effort to address the dreadful cholera epidemics that were plaguing Europe at the time. His water filtration technique was a tremendously significant improvement in water sanitation practices and made great strides in public health. Historians have claimed that this one contribution was Koch's most important one in terms of reaching the general populace worldwide.

Koch participated in developing the notion in 1902 of the so-called carrier concept. The idea held that asymptomatic individuals could carry disease-causing microbes without showing any signs of disease but nonetheless able to transmit pathogenic microbes to other unsuspecting individuals.

Certainly, each of these individual contributions made by Koch in their own right would constitute a singularly significant advance in the history of the biological, biomedical and microbiological sciences. In this light, Dr. Koch is undoubtedly on par with the great Dr. Pasteur.

2) Apparently, he also experimentally tested Pasteur's germ theory of disease. Do you know how he did this?

In the 1850s, Louis Pasteur formulated his theory that microbes (germs) are a cause of disease, a type of ailment known as an infectious disease. Koch was the first scientific investigator to put the germ-disease notion to the test. Koch used a set of investigative principles outlined by Dr. Edwin Klebs, principles of which were later to be named after Koch, who elegantly put the postulates to good use in discovering that a bacterium caused a disease known as the anthrax. In so doing, Koch validated Pasteur's enduring hypothesis.

Koch started his anthrax experiments with diseased sheep. He took blood samples from sheep that had perished from their anthrax malady and looked at the dead sheep blood under the microscope. In the diseased sheep blood, Koch saw what could be described today as bacterial endospores. This apparently had been the very first time in history that any human had visualized such microbial structures and had recorded the event, for posterity.

He then injected experimental laboratory rabbits with the blood from the diseased sheep. The test rabbits that were inoculated with the diseased sheep blood quickly developed the anthrax and died within a day or two. Koch then excised ear and skin tissue from the dead rabbits and preserved them in alcohol.

These and other bacteria-harboring rabbit tissues were then used by Koch to surgically inoculate the eyes of new test rabbits. The newly inoculated rabbits quickly succumbed to the anthrax and died. During the rabbit necropsy (animal autopsy), Koch found their tissues teeming with the bacteria. He noted their characteristic bacillus (rod-like) shapes.

Next, Koch undertook tremendously laborious efforts to cultivate the bacteria that he had derived from the tissue of the dead rabbits. In this respect, Koch was said to have been fortunate in that he had injected the rabbits' eyes—it helped to provide an astute insight—to use rabbit eye cornea fluid as a growth medium in order to grow the bacteria in the laboratory as a culture. In so doing, Koch succeeded where others had failed, as Koch was able to successfully culture and isolate his anthrax bacteria.

It is now at this point in his protocol where Koch performed his coup de grâce. He injected laboratory test mice with his purified isolated bacteria. The mice quickly died of anthrax. Subsequent necropsy showed that the dead mice were laden with the same bacteria that he had seen before.

In his 1876 publication, Koch brazenly announced that the bacillus-shaped bacterium was the contagion of anthrax. He even named the bacteria as such, *Bacillus anthracis*, to quell any lingering doubts or suspicions regarding the validity of the germ cause of disease.

3) Apparently, in the year 1877 Koch published a very important paper on the photographing of bacteria. Why was this important, even way back then?

Koch's 1877 paper, published in the journal *Beiträge zur Biologie der Pflanzen*, which loosely translates to '*Contributions to the Biology of Plants*,' was the very first photographic depiction of a microbe. Indeed, it was the first time in history that the general public saw what these microbes actually looked like. Prior to this, starting with Leeuwenhoek, the scientific journals merely had only drawings of the various microbes.

The 1877 publication forever established the new field of photomicroscopy. In the article, Koch revealed new protocols for fixing bacteria onto microscope slides, staining the bacteria to see them better, and amazingly, his descriptions of technique were so well detailed that others could readily adopt them today, for everyone to use, just as well as Koch had done in the 19th century.

In the keynote paper, Koch showed photographs of his infamous anthrax bacilli and their equally infamous endospores. Fascinatingly, he included elegant images of helical shaped bacteria, called spirochetes and spirilla. The twisted shapes of these bacteria are easily discernable. The 1877 black and white photographs of Koch's microbes could easily pass for modern photographic images. It was, however, painstaking work to acquire images that Koch considered good enough for publication.

The 1877 Koch article is important even to this day as for not only establishing the discipline of photomicroscopy but also for establishing a new and enduring means of universal scientific communication, in both research and educational endeavors. Textbooks and journals alike regularly include photographs of microbes. It all began with Koch.

4) For his discoveries in regard to tuberculosis, he received the Nobel Prize for Physiology or Medicine in 1905. Have we finally eliminated tuberculosis, or does it still exist in certain parts of the world?

Robert Koch was the first bacteriologist and one of the first microbiologists to have garnered the Nobel Prize. His primary contribution in this regard was that he had discovered the causative bacteriological agent of tuberculosis (TB).

TB is a terribly dreadful infectious ailment of the lungs, i.e., a bacterial pneumonia, characterized by a persistent cough. TB has been said to 'consume' a patient to death. In historical terms, therefore, TB has been called 'The Great Consumption.' The consumptives in Koch's time died a slow and agonizing death.

Koch published his history-making TB findings in 1882. The TB-causing bacterium is known as *Mycobacterium tuberculosis*.

Koch's discovery was important at the time because it identified a new target, *M. tuberculosis*, for the discovery of new antimicrobial therapies against the TB. Indeed, such anti-tuberculosis chemotherapy came to fruition during the 1940s and early 1950s.

Sadly, however, not only have we not eliminated tuberculosis, its incidence and prevalence have only gotten worse. The TB vaccine, known as the Bacille Calmette-Guerin (BCG), is unfortunately not terribly effective in preventing the TB disease.

Worse, however, was the emergence of new devious strains of the tubercle bacterium that were resistant to the anti-tuberculosis medicines used to treat the TB. Not only did the new medicines not work to kill the bacteria in the consumptives, newer strains became multidrug resistant. Furthermore, in a cruel and deadly new manifestation, recently emerged M. tuberculosis strains, though presently rare, are virtually untreatable in that they are extensively drug-resistant TB (XDR) variants. The XDR TB are insidious and deadly.

5) I think Koch's postulates are so important, that we should reproduce them here for scientists and for you to comment briefly on each.

1. The organism must always be present, in every case of the disease.

What Koch is referring to here is that in every examined dead animal or living host with active disease, the microbe must be observed as having been present in each case of the diseased victim. In principle, there should be no exceptions to the postulate. Sampling of the patient should show that the microbes are present.

2. The organism must be isolated from a host containing the disease and grown in pure culture.

In this scenario, Koch is saying that from the diseased host, the experimenter must be able to sample the host, acquire the microbe, and then successfully culture the sampled microbe in a pure form. There should be no other microbial contamination in the grown culture.

This second postulate of Koch's has one drawback: some microbes are simply uncultivable in the laboratory. Although such microbes may grow freely in their host victims, they are unable or unwilling to grow outside of the host in artificial environments, like in a research laboratory setting in a Petri dish, a culture flask or a test-tube. These types of microbes cannot be grown on any known artificial culture medium yet devised by microbiological investigators. The microbiologists frequently refer to these disappointing microorganisms as viable but non-culturable (VBNC) agents. In the case of the VBNCs, modern experimentalists may sample the metagenome in order to satisfy this postulate—if the VBNC microbe cannot be grown in culture, then the presence of their DNA may suffice for now, at least in some cases.

3. Samples of the organism taken from pure culture must cause the same disease when inoculated into a healthy, susceptible animal in the laboratory.

Here, Koch is referring to the requirement that if a microbe indeed causes a disease, then one should be able to inject the pure microbe into a disease-free host and watch the inoculated host get the disease. The new host should be initially free of any such disease prior to the microbe inoculation, of course. The host should then acquire the very same disease that had been originally observed in postulate no. 1.

This particular postulate has been problematic for microbes that infect only humans. For example, poliovirus and human immunodeficiency virus (HIV) infect only humans and apparently no other kind of living host, as well as they do in humans. No animal model may exist for this postulate in such cases. Thus, it may be rather difficult, ethically speaking, to inoculate humans in this way just for the sole purpose of observing the disease in new human hosts. The investigator may attempt, sometimes successfully, to use human cells in culture for this purpose.

4. The organism must be isolated from the inoculated animal and must be identified as the same original organism first isolated from the originally diseased host.

In the fourth and final postulate requirement, Koch is rightly insisting that after the healthy animal inoculation and disease-causing step, performed in satisfying postulate no. 3, the host with the newly acquired disease should harbor the same microbe. The microbe that is sampled from these inoculated hosts must be identical to the very first microbe that was purified in the first place.

After the fourth postulate is satisfied, the clinical investigator can state, with a fair degree of certainty, that the postulate-satisfying microbe is the etiological agent of the infectious disease studied.

In terms of science, how are these still relevant today?

Amazingly, Koch's postulates are still followed to this day. In fact, if just about any clinician or basic researcher would like to claim that a microorganism causes an infectious disease, they shall be forever required to experimentally satisfy each of Koch's postulates.

I am convinced that Koch's postulates will most certainly continue to be in vogue well into the far away future.

6) One key thing at least in my mind was his efforts to develop new methods for staining bacteria, which obviously made them more easily visible to the naked eye and which helped to identify them and perhaps categorize them. Am I correct here?

Koch and his laboratory personnel worked tirelessly in this regard. With a few notable exceptions (like the giant prokaryote called *Epulopiscium fishelsoni*), bacteria cannot be seen with the naked eye. Most bacteria are too tiny to be seen by an unaided eye, unless using a microscope. Further, many microbes are difficult to study without any stain on the microbial glass slide when viewed under microscope. Without stains, the investigators could see only ghosts of empty cells, making it terribly difficult to study such microbes effectively. Koch tried many new chemicals to stain their microbial isolates in order to see them better under the microscope lens. They used aniline dyes, which are used to stain wood, to stain bacteria. The approach worked really well as Koch could easily visualize bacteria.

They also used a chemical called eosin, still in use today, to stain eukaryotic white blood cells. Other chemicals used by Koch include safranin (Gram's counter stain) and fuchsin, still used today to stain bacteria.

Koch even tried to visualize bacterial flagella, which are molecular propeller-like motors that permit bacteria to swim, but sadly, this attempt was not successful. Later investigators, however, were indeed successful in staining bacterial flagella.

7) Koch presented the notion that bacilli which caused human and bovine tuberculosis were distinctly different and separate. Although this was initially mocked, it seems that again, he was correct. Why is this relevant and important?

Indeed, in 1901, Koch had made the startling announcement that the causative agents of human versus bovine TB were two entirely different species of bacteria. That was, however, after the occasion in 1882 in which he had made the equally startling announcement (more like an incorrect opinion) that the human and bovine TB bacteria were of the same species.

Changing Koch's mind about the distinctiveness between human versus bovine TB took years to reach the correct conclusion. He had taken notice of the data suggesting that the human and cattle diseases had different symptomologies. He further noted that individuals, primarily children, who consumed milk from tuberculate cattle hardly ever acquired the pulmonary TB. Most convincingly of all, the bacteria that had been isolated from both human pulmonary and bovine gastrointestinal subjects had distinctive bacteriological properties. Indeed, it was experimentally confirmed that human TB was caused by *M. tuberculosis* and that bovine TB was caused by *M. bovis*. That is to say, the two diseases involved completely different microbial species.

In attempting to clear up the controversy in 1901 by stating decisively (and correctly) that the human and bovine TB bacteria were actually different from each other, he then went on to make an even more shocking (and inaccurate) pronouncement and started a new controversy.

He had incorrectly proclaimed that bovine TB prevention measures, like milk pasteurization, were not necessary because, as he further incorrectly stated, the bovine TB bacillus was not transmissible to humans; he was referring, however, to the notion that patients could rarely get the pulmonary TB disease from the bovine bacillus in milk. His critics rose to also state correctly that the bovine bacillus still caused gastrointestinal TB.

The controversy came to a head in September of 1908 at an international scientific symposium, held at the famous and still standing Willard Hotel, in Washington, D.C. With both sides at an apparent standstill, it was agreed to hold an additional closed committee meeting during the symposium, in order to settle the issue for the last time.

The disagreement was essentially as follows. There were those who were in favor of the idea that bovine TB bacillus transmission to humans WAS NOT possible (pro-Koch), and there were those in favor of the idea that the bovine TB bacillus transmission to humans WAS possible (anti-Koch).

A stenographer had been called to record every word spoken during this special closed meeting in 1908, and the discourse was published later for all posterity in the *Journal of the American Medical Association*. The meeting was, not surprisingly, contentious. In the end, it was resolved that both forms of TB, bovine and human, were transmissible.

The resolution was a public and far-reaching humiliation for Koch, as both Germany and the U.S. had not advocated milk pasteurization because of his earlier pronouncements—prior to this symposium and its resolution, Koch's every word, it seemed, had been given great weight. Shortly after the 1908 symposium, the US adopted the widespread practice of pasteurizing milk—at the precise temperatures that would suffice to kill both forms of the TB bacilli. Germany apparently followed suit a little while later.

8) What have I neglected to ask about this very hard working scientist?

Astonishingly, Robert Koch and Louis Pasteur, two of the most famous scientific investigators of all time, did not get along! Although their first meeting in 1881 at a scientific conference in London on a face-to-face basis was an amicable one, it was sadly to be their last friendly one.

Since neither investigator had spoken the other's native language very well, they had used translators. In one unfortunate incident, however, in which Pasteur was delivering a symposium keynote speech, he used a French term ('recueil') that he had meant to describe

Koch's compilation of German works; but Koch's amateur translator mistook the French word to mean the German word, orgueil, instead, which meant 'pride.'

Koch was profoundly insulted by the inference of the so-called 'German pride.' Without waiting for confirmation, Koch rose in front of the audience to denounce the perceived insult whereupon the surprised Pasteur simply (and seemingly wantonly) waved off Koch.

Unable to speak efficaciously to each other because of the language barrier, each prominent investigator chose to air their differences in the scientific literature, for all to read. In the literature, they were repeatedly personal in their stinging remarks about each other. Their students took on the fight and were even ruder in their criticisms. The rift between these two greats lasted years and ended only with the death of Pasteur.

Chapter 11

WALTER REED

1) Dr. Varela, I would say that almost once a week, we hear on the news about Walter Reed Army Hospital. What do we know about this individual and their early life?

I think that it is quite appropriate that you would inquire about Walter Reed. In our list of amazing scientists, he is the first American we have considered in this series. Reed is famous for having been the first investigator to correctly establish that the dreaded Yellow Fever is transmitted to humans by the bites of mosquitoes. Having also served in the military, Reed has been honored by the naming of his army's hospital after him: Walter Reed Army Hospital. It is a well-known institution and the chief healthcare facility for U.S. presidents.

Walter Reed was born on the 13th of September in 1851 in the county of Gloucester, in Virginia. Walter's father was Lemuel Reed, a clergyman in the Methodist church, and his mother was Pharaba White, Lemuel's first wife. Walter had four older siblings and was the youngest of the family. Unfortunately, his mother died in 1865, when Reed was only 14 years of age. According to biographers, Reed was devastated by the death of his mother, and took refuge in study. Reed's father remarried a wealthy widow, Mary Catherine Byrd Kyle. The new marriage provided much needed support and stability to the family, their finances, and to the education of Walter Reed.

2) Where and when was he educated?

Because of his father's ministry in the Methodist church, Walter Reed's family had to move from parish to parish every couple of years. Thus, Reed's early education was considered sporadic and haphazard. Stability in his educational endeavors took hold almost immediately after the American Civil War ended when the family arrived in

Charlottesville, VA. Reed initially attended Abbot School, a day school run by a Civil War veteran.

At the age of 15, Reed was eligible to attend the University of Virginia, established by Thomas Jefferson, where, unable to afford the M.A. degree, he instead applied for entry into medical school and later earned his M.D. in 1869 at the age of 17 years. Like many physicians of his time, an undergraduate education was not required for entrance to medical school. Unfortunately, at the time, medical students attending the University in Virginia received little, if any, training with actual patients.

Interestingly, with his M.D. in hand, Reed then moved to New York University, where he then entered another medical school, this time Bellevue Hospital Medical College, where he was actually allowed to work on patients. Furthermore, the New York medical school was the first in the U.S. to teach anatomy using human cadavers—although the providers, known as the "resurrectionists," of these cadavers were evidently of dubious origins—having stolen the bodies of executed prisoners or of the destitute, for dissection. The Bellevue Hospital Medical College claims a series of other firsts for the U.S., such as the first to use hypodermic needles and syringes, plus the first to conduct outpatient practices, first to open an ER (called an "emergency pavilion"), and the first to establish a school for nursing.

Other medical subjects at the Bellevue institution included chemistry, pharmacy, medicine, physiology, and surgery—the latter subject of which was taught by Prof. Lawrence Cabell, who was an advocate of Listerism. Despite this, infections such as the dread puerperal fever and erysipelas, were rampant throughout the obstetrics department, and sepsis was widespread in the surgery department.

Listerism had not yet taken hold during the time that Reed was enrolled. Reed was exposed to the harsh realities evident in the patients at the urban hospital, such as poverty, overcrowding, poor sanitation, exploitation, suffering, disease, and despair, especially amongst the young and in the newly arrived immigrants looking for a better life in New York and in the U.S. The plight of the patients at Bellevue had a profound and long-lasting effect upon Reed and played an important role in Reed's later efforts to address serious yellow fever epidemics.

In 1870, Reed graduated near the top of his class and took an unprecedented second M.D. degree. Incidentally, William Worrall Mayo, of the future Mayo Clinic, was a classmate of Reed's. Next, Dr. Reed began a series of internships and residencies in New York, including Kings County Hospital and Brooklyn City Hospital, where he learned about the potential opportunities for preventative measures towards combating infectious disease epidemics.

During this time, one prominent and influential figure regarding these preventative measures against contagions was Dr. A.N. Bell, who once claimed to have stopped a yellow fever outbreak with steam as the disinfectant. Sadly, Bell later became a vocal critic at a scientific conference where Reed presented his notion of yellow fever disease transmission by mosquitoes. In 1875, Reed joined the military becoming a U.S. Army surgeon with the medical corps at the rank of Major, and in 1893 Major Reed became a professor of bacteriology at the School of Medicine housed in George Washington University, located just a few blocks from the White House, in Washington, DC.

3) Now, I suspect that the vast majority of us, at some time or another have been bitten by a mosquito- and it itches for a brief period of time. Now, why is the mosquito a somewhat dangerous insect and what is the connection between the mosquito and "yellow fever" (and what exactly IS yellow fever and what are the symptoms?)

Simply put, some of the various mosquitoes can be vectors of certain infectious disease transmissions to humans. For instance, *Aedes aegypti* can transmit the Yellow Fever, Dengue, and Zika viruses; while *Culex pipiens* transmits the West Nile virus, and *Anopheles* spp. will transmit the malarial agents. Incidentally, Dr. Reed was not the first to demonstrate this particular mode of disease transmission. Rather, Reed was the first to experimentally and definitively put this hypothesis to the test with the yellow fever transmission to humans by these types of mosquitoes.

Furthermore, that mosquitoes could transmit disease at all was first shown in the 1890s by Ronald Ross, Amico Bignami, Giovanni Battista Grassi, and Giuseppe Bastianelli, when they provided the first experimental evidence that malaria could be transferred to humans by the bite of a mosquito. Prior to this discovery, it had been postulated that "bad air," the miasma, was responsible for the malaria, i.e., mal (bad) aria (air). The microbial causative agent of malaria that's transmitted by the mosquitoes is a tiny protozoan worm, the size of a red blood cell, called *Plasmodium falciparum*. Other malaria-causing protozoans include *P. ovale*, *P. malariae*, and *P. vivax*.

Yellow fever, on the other hand, is an infectious disease caused by a microbe called a virus, and like malaria, the yellow fever illness is severe. Shortly after the virus is transmitted to the patient by the bite of a virus-carrying mosquito, symptoms emerge, such as the characteristic fever, headache, nausea and vomiting. The patient may experience chills, tiredness, loss of appetite, and soreness in the muscles.

The patient might also be delirious from the fever. Damage to various organs can occur, such as to the gastrointestinal tract, liver, kidneys, and the heart, leading in many cases to shock—a serious and potentially lethal situation. Internal bleeding, called hemorrhaging, may occur, and if blood enters the intestinal tract, it produces the so-called "black vomit" that was so evident in Reed's yellow fever patients. Consequently, the Spanish referred to the disease as the "vomito negro."

Further, damage to the liver produced jaundice, which manifested itself in the visibly intense yellow color in the patient's skin, including the eyes. With the patients in their beds, all that could be frequently seen were their bright yellow faces. The yellow fever was often referred to as the "yellow plague," primarily for this reason. Additionally, the illness was called the "yellow jack" because at the time quarantined ships harboring it would fly a yellow flag instead of the union jack to indicate its presence. Ships flying the yellow jack flag were considered infectious and not permitted entry into onshore ports.

In Reed's day, yellow fever ravaged both the North and South Americas. With a 20 to 50% mortality and during mosquito season, it killed thousands annually. In fact, it is said that during the 1898 Spanish-American War more people were killed by yellow fever than by bullets. Today, it is virtually eliminated from North and Latin Americas, but many thousands worldwide die annually from yellow fever.

The yellow fever virus itself consists of only a single piece of an RNA genome molecule surrounded by a single protein coat, called a capsid, which in turn is surrounded by a membrane, called an envelope. The viral RNA genome codes for only a handful of proteins, mainly for the viral capsid, but also for assembly and processing of the mature yellow fever virus. The envelope is stolen by the immature virus as it escapes the human cells it infects. The yellow fever virus belongs to the genus *Flavivirus* and the family Flaviviridae.

Incidentally, this viral family harbors also the Zika virus, the Dengue virus, the West Nile virus (each of which are mosquito-borne) and the Hepatitis C virus (not transmitted by mosquito vectors), all viruses of which are important pathogens in their own right.

It may interest the reader to know that Dr. Reed actually was not the first person to propose the idea that mosquitoes passed along the yellow fever. That credit goes to Dr. Carlos Finlay of Havana, Cuba, who first postulated the idea. In fact, Dr. Finlay even published evidence for the mosquito-yellow fever connection. The problem was no one believed him, as many in the field criticized Finlay's work.

Prior to the establishment of the mosquito-yellow fever link, it had been mistakenly thought that the yellow plague could be prevented if not treated; for example, it was thought that clothes and bedding of yellow fever patients could be fumigated or even burned; it was further advised not to shake hands with patients. It was even believed that homes of patients should be burned to the ground.

Dr. Reed enters the picture here as the leader of the so-called Yellow Fever Commission, sent to Cuba in 1900 to investigate the outbreak there in Quemados. First, invoking Koch's Postulates, Reed and his team tested the yellow fever patients, and no bacteria could be found to cause the disease. It was considered a serious failure.

The Commission paid a visit to Dr. Finlay and listened to his mosquito theory. He even provided members of the Commission with mosquito eggs, which Reed's colleagues hatched in the laboratory, and searched for bacteria—finding none that passed or satisfied Koch's postulates.

In support of Finlay's theory, however, Reed noticed that the nurses who treated the yellow fever patients in the ward never got the disease themselves—even though the nurses were in constant contact with bloodied, soiled bedding and clothes of the patients. Considering also the air-borne mode and the time it took for one patient in a household to get the yellow fever, recover and then weeks later see other members of the same house then get the disease, Reed thought that perhaps the contagion needed time to develop, in the air, perhaps in an insect, for two weeks. Suddenly, Finlay's mosquito idea made sense.

A second experiment was, thus, conducted by Reed and his Commission. It entailed mosquitoes biting yellow fever patients and then letting these mosquitoes bite an unexposed person. Calling for volunteers in his Commission, several people rose to the occasion and literally took a bite with yellow fever-laden mosquitoes, for the cause. All volunteers of the Commission got the yellow fever, but fortunately survived.

One unfortunate mishap occurred, however, with one of these survivors, a one Dr. Jesse Lazear, who later had been bitten again with a presumably stray yellow fever mosquito back in the lab—and a mere twelve days later, on September 25, 1901, he was dead from the yellow fever. Some argue that it was an accidental bite, whereas others argue it was a deliberate act by Lazear, in the name of science. To this day, it is unclear which the case is.

Nonetheless, we do know that the experiment definitively demonstrated that yellow fever was transmissible by mosquitoes.

A "dirty" third experiment was conducted by Reed and his Commission. The question remained whether other modes of transmission were additionally responsible for the transmission of the dreaded yellow fever contagion. Perhaps yellow fever was, for instance, blood-borne. Thus, with additional volunteers at hand, they opened new boxes delivered to them—they contained pillows, pajamas and sheets soiled with blood and black vomit, all items of which had belonged to patients who had just died of yellow fever.

The volunteers actually wore the same uncleaned bedclothes and slept on the uncleaned pillows of the patients dead from the yellow fever disease—for almost three weeks! These volunteers did not get the yellow plague, further solidifying the mosquito-yellow fever link.

A new criticism, however, provoked another experiment. Critics within the Commission speculated whether the human subjects in Reed's experiment weren't already immune to yellow fever in the first place—perhaps this prior "immunity" is why they did not come down with the disease during the last experiment. Therefore, the new experiment entailed giving two of the same and potentially immune volunteers actual cases of the illness with yellow fever-laden blood—this time in a clean environment—sans the bloody pillows and clothes, etc. The volunteers, both of whom had not gotten the yellow fever in the previous dirty experiment, now got full-blown cases of it this time in the clean experiment. Both volunteer patients eventually recovered from their ordeal.

It was now clear that mosquitoes were to be seriously considered the transmitters of the dreaded yellow fever.

4) I may be wrong, and I would certainly apologize, but did Walter Reed have anything to do with the Panama Canal? (or perhaps my history is off?)

As you may have deduced when you formulated this inquiry, the building of the Panama Canal was indeed fraught with the very serious problem of the yellow fever, which was endemic during this time period, in the early 1900s. Disease was delaying progress. Thousands of their workers were reported to have died from the yellow fever and the malaria, which as you'll recall, is also transmitted by mosquitoes.

The Canal was completed in 1914, probably earlier than had been anticipated, as a result of Reed's famous discovery. You see, after the historical Reed experiment in Cuba, the Panama Canal builders immediately commenced mosquito control, with mosquito netting, screens on windows, and spraying with chemicals that killed the insect vectors. The yellow fever epidemics subsided, and, thus, the Canal was finally finished. Hence, Reed most certainly had a hand, however indirectly, in facilitating the eventual completion of the Panama Canal.

5) His legacy obviously lives on today, but what would YOU say are his outstanding accomplishments and achievements?

Unfortunately, Dr. Reed died unexpectedly from a complication of a ruptured appendix during appendicitis, a serious condition at the time called peritonitis, on the 22nd of November in 1902, just shortly after he and his Commission had experimentally demonstrated that mosquitoes carry the yellow fever virus and transmit the microbe to human beings. Regrettably, any further accomplishments by Reed were to be cut short. Just prior to his death, there had been talk of Reed becoming the new U.S. Surgeon General. Obviously, this did not happen.

I think one achievement, alluded to above, deserves a mention here, as well, and that is that Reed became a full-fledged physician at such an early age, in his teens. Reed apparently still holds the record of being the youngest student ever to attain an M.D. degree at the University of Virginia medical school.

One important accomplishment was that Reed had ruled out many other possibilities for yellow fever disease transmission. This turned out to be an important contribution in later efforts to reduce morbidity and mortality of yellow fever.

An important contribution by Reed was that he had demonstrated in 1901 that the causative agent of yellow fever was a filterable agent, thus showing that it was a virus. This was the first report of a disease in humans being caused by a virus.

There were, at the time of his discovery, a plethora of critics and criticisms regarding a disbelief in the mosquito-yellow fever connection. The theory, nevertheless, has withstood the test of time. It has held up to greater scrutiny by many investigators throughout the years after Reed's death. Thus, I think one outstanding achievement is that he was correct early on.

Related to this achievement, too, is the fact that Reed was always quick and eager to give credit to Finlay for having been first to postulate the mosquito idea, both in scientific presentations and in his publications; Reed merely tested Finlay's idea with carefully controlled experiments. This relationship is reminiscent of Pasteur and Koch; Pasteur formulated the germ theory of disease, and Koch experimentally put it to the test. Each of these theory formulations and the evidence generated for them are significant milestones in medical and biomedical history.

Another significant criticism is that Reed used human subjects, putting them in grave danger with the yellow fever mosquito bites and the blood injections. Perhaps he might have considered for his experiments using lower primates, but it was not clear then that they were good yellow fever hosts (they are but not for getting the disease). To his credit, however, Reed had actually attained an informed consent of sorts from his volunteers, who seemingly knew the purposes of the experiments and all of the associated risks, including that of death. The informed consent practice was not widely adopted till well into the 20th century, and Reed practiced it long before it became standard.

6) What have I neglected to ask?

I think it's fascinating to know that certain prominent and historical figures suffered from the yellow fever and to understand how they dealt with it. For instance, in 1793 Abigail Adams, wife of President John Adams, had a severe case, almost dying of it as described in a letter she wrote, detailing its effects. That same year, Abigail recorded in a letter to her sister on the outbreak of the yellow fever in Philadelphia, where thousands had died from the pestilence. Thomas Jefferson wrote about the efforts taken by him, President George Washington, and his wife Martha, to prevent the appearance of panic, such as quietly fleeing the city of Philadelphia to Monticello and Mount Vernon, respectively. Similarly, both Alexander Hamilton and his wife, Elizabeth Schuyler (Eliza), endured critical cases of the yellow fever, both of whom nearly died from its effects.

Importantly, because of periodic outbreaks, yellow fever affected historical events at the time, such as delaying the secret meetings of the patriots for the purpose of planning the American Revolution—the secret meetings had to wait until the yellow fever season was over—usually after the cold weather set in when, unbeknownst to these historical figures, the mosquitoes would cease in their transmission activity. Therefore, I believe that had it not been for the yellow fever (and perhaps the smallpox) having delayed important events in U.S. history, Americans would be celebrating another year besides 1776.

Our readers may also appreciate that the causative agent, the yellow fever virus described above in detail, was not isolated until 1927, many years after Reed's death. A useful vaccine was shortly thereafter developed. Since then, the virus has been extensively characterized, and much is known today, for instance, about its molecular structure, its mechanism of cellular infection, its pathogenic mechanism, and its gene expression regulation. Despite these advancements, yellow fever remains a significant problem in tropical and sub-tropical areas of the globe, and treatment is still limited to simply addressing the severest of the symptoms. The vector may still be found throughout much of the southern U.S. region. Thus, if yellow fever were to become re-established in these and other areas, it could cause serious medical problems. Thus, prevention today has been key, with, for instance, vector control and vaccination.

Chapter 12

RICHARD JULIUS PETRI

1) Dr. Varela, most scientists have heard of the Petri Dish, but many have no background knowledge as to who developed it, or when. So, let's set the record straight- Who was Richard Julius Petri, and where was he born and when?

Richard Julius Petri was born on May 31, 1852, in Barmen, Germany. He became a medical doctor and later a bacteriologist, as an assistant to the famous Dr. Robert Koch. Dr. Petri was to become well known as the so-called inventor of the Petri dish, a small glass container used for growing microbes, such as bacteria, and which is still in use to this day by modern bacteriologists, molecular biologists and scientists studying the many fields of the microbiological sciences. Petri dishes are also used to study other microbes, such as the fungi, algae, and viruses. I've even seen biologists use the Petri dishes to grow other organisms, such as flies and plants.

2) What do we know about his early education and childhood experiences?

Petri received an elementary and high school education in his hometown of Barmen, Germany. Details of Petri's childhood experiences, however, are scant.

3) Where did he go to university or college and what do we know about his early work?

Details of Petri's undergraduate education are unclear. It is known that Dr. Petri took his M.D. degree in 1876 from a military institute for physicians, called the Kaiser Wilhelm Academy, in Berlin, Germany. The learning mode of his education there used the apprenticeship format. During this period, Dr. Petri continued his education at the teaching hospital called Charité, in Berlin. After earning his doctorate, Dr. Petri became a military physician between 1876 and 1882. But in 1877, Dr. Petri also began conducting scientific research till about 1879, at the so-called Imperial Health Office, where he became a laboratory assistant to the famous Dr. Robert Koch and where he made is famous Petri dish invention.

4) Now, on to the actual Petri Dish- a staple of scientific work. When did it first come about or when was it first used, and for what purpose?

The state of research prior to the publication of Petri's paper and leading to his Petri dish invention, dealt with the attempts by investigators in the field to purify microbes, called bacteria, when growing them in the laboratory, a process called culturing. Early methods employed by Prof. Louis Pasteur and others to culture bacteria in pure form involved using a liquid medium called broth. While such cultures had largely pure species of bacteria, one was never quite certain that these liquid broth cultures were not contaminated with unwanted bacteria. Dr. Joseph Lister was first to demonstrate a pure bacterial culture using a dilution technique, in 1878. Using a solid medium, however, demonstrated a better way to purify bacteria because one could easily see a bacterial colony, and by successively transferring these colonies to new solid material, one could be more certain that the bacteria were pure. The first good solid medium, developed by Joseph Schroeter in 1875, was a potato slice, and good colonies could be seen on the slices. Unfortunately, not all bacteria grew on these potato slices.

Dr. Koch enters the picture by using gelatin, the same material used today to make Jello, to culture bacteria and to see their colonies. His work became so successful that he was able to develop the so-called 'Koch's plate technique' in which he streaked bacteria onto his gelatin plates by diluting the bacteria with each successive streak, a technique known today as the streak dilution inoculation method.

One interesting problem emerged, however, with the use of gelatin; it tended to melt in the incubators, making it difficult to use effectively. A more interesting solution became known when the spouse of Walter Hesse, a lab assistant in Koch's laboratory, named Angelina "Fanny" Hesse, suggested they use agar in their plating technique, instead of the gelatin. This simple medium exchange permanently altered scientific history, as all microbiologists today still use agar and Koch's plating method.

However, at the time Koch and his lab assistants, including Petri, covered the plates with large bell jars during incubation of the bacteria on solid agar containers. The bell jar method was cumbersome and not entirely amenable to the eradication of contaminating bacteria landing on the agar from the air trapped within the bell jar. It was a difficult process to carryout.

This is where Petri made his extremely successful discovery. He merely adapted the plate technique of Koch by providing a small lightweight cover dish, instead of the larger bell jar method, to cover the bottom dish containing the agar. In this way, the resultant plates were easier to make, easier to use, and most of all, lessened the possibilities for contamination. The scientific paper detailing Petri's dish method was published in 1887. It was a short and simple two-page article titled "*A minor modification of the plating technique of Koch*."

5) When was it mass produced and how was it received?

The use of Petri's dish became widespread amongst bacteriologists of the day almost immediately after the publication of Petri's seminal paper, in 1887. Koch himself used his plating method along with Petri's efficient dish to culture and isolate bacteria that caused tuberculosis, earning Koch the Nobel Prize in physiology or medicine. Dr. Alexander Fleming used Petri's dishes to discover his famous antibiotic penicillin, earning him a Nobel, too.

These early Petri dishes were made of glass. Mass production of the Petri dish took off during the early 1900s, when glass makers like Pyrex and Corning fine-tuned their production methods. The glass Petri dishes had to be sterilized before use and cleaned and sterilized again after use. As such, they were routinely reused when experiments were conducted in the laboratory. Later Petri dishes were, and still are, made of cheaper lightweight disposable plastic. Nonetheless, whatever material the Petri dishes are made of, they are widely used throughout all scientific and educational venues today.

6) What other events or discoveries were attributed to him?

It has been reported that Dr. Petri published over 150 papers during his career. In addition to inventing the famous Petri dish, Dr. Petri developed a method for growing and purifying bacteria in pure culture by using slanted agar in test tubes and alternating their transfer between the slants and Petri dishes. The slant culturing method is still in use today.

In 1889, Dr. Petri published a paper pointing out the potential dangers inherent in using carbonated soda in furnaces. Dr. Petri also published a paper in 1893 pointing out how the contagion of tuberculosis moved within populations by train. Dr. Petri published a paper in 1987 about testing the qualities of milk and butter. Dr. Petri also published reviews, each dealing with specialized topics, such as bacteriological research methods, industrial hygiene, and the microscope.

7) How did he spend the last years of his life, and what have I neglected to ask about him?

After working in Dr. Koch's laboratory, Dr. Petri went on to lead a tuberculosis sanatorium, until 1885. He subsequently became the director of the Museum of Hygiene, located in Berlin, until 1889, when he then returned to the Imperial Health Office to become its director. After his retirement in 1900, the title of Privy Councillor (Geheimer Regierungsrat) was bestowed upon him. Petri died in Germany at the age of 69 in 1921.

Chapter 13

CHRISTIAN GRAM

1) Dr. Varela, we often hear the term "Gram staining technique" in science and microbiology and the like. What exactly does this mean and why is it important?

The Gram stain technique has been around since Dr. Hans Christian Gram developed it during the mid-1880s. No doubt you have heard about it often because Gram's technique is routinely taught in all Microbiology laboratory courses and even in good freshmen Biology lab courses. Furthermore, the Gram stain is regularly used in clinical medicine and medical diagnostic laboratories.

It is a staple procedure in each of the infectious disease specialties. In fact, the Gram stain is frequently the very first laboratory test that's conducted when a new species of bacteria is discovered and characterized further. The Gram status of all bacteria is used in taxonomic and systematics circles to classify all extant and newly discovered bacteria. Many bacteria are classified as Gram-positive or Gram-negative.

If a bacterium is denoted as Gram-positive, it means that the chemical stain, called crystal violet, stays with the bacterium when washed; on the other hand if a bacterium is Gram-negative, it means that the stain comes off the cell after washing. Depending on the nature of the bacterial species being studied, the cell walls that cover a Gram-positive versus a Gram-negative bacterial cell will have distinctive biochemical and cellular structures within their cell wall coverings. These different biological properties of bacteria are used to classify them and to differentiate between various bacterial species, and these properties are often focuses of both basic and applied scientific research areas.

2) So, who exactly was Christian Gram, and where was he born and what do we know about his early life?

Hans Christian Joachim Gram was born in Copenhagen, Denmark, on September 13, 1853, to parents Frederick Terkel Julius Gram, a law professor, and Louise Christiane Roulund.

3) Where did he attend university and how did he get involved in science and research?

Gram was an undergraduate at the University of Copenhagen, in Denmark, where he primarily focused his studies in the field of botany, which is the study of plants. Gram reportedly took his undergraduate degree in 1871. This educational activity introduced Gram to the important scientific fields of microscopy, as well as in pharmacology. He then became a botany assistant under the famous Danish professor of zoology, Japetus Steenstrup. Gram went onto medical school and took his M.D. degree in 1878; he then earned his doctorate in 1883 studying erythrocytes under the supervision of Prof. Steenstrup. While travelling about Europe, Dr. Gram took up a new position in 1883 with Dr. Carl Friedländer, a bacteriological pathologist housed in Berlin, Germany. It is in Friedländer's laboratory that Dr. Gram developed his now famous Gram staining procedure.

4) This "gram staining technique for bacteria" - how did it come about and what is its relevance in terms of bacteria? And does it apply to viruses also?

The story of Gram's discovery began with a controversy. At the time, there had been a disagreement between Gram's supervisor, Dr. Friedländer, and Dr. Albert Fraenkel, a German physician. Dr. Friedländer had isolated a bacterium from the lung of a patient who had died of pneumonia. While Dr. Fraenkel had also isolated a bacterium, but from another patient, this one having died from lobar pneumonia. Fraenkel was convinced that his bacterium and that of Friedländer's were one and the same, while Friedländer was equally convinced that the two bacterial isolates were completely different from each other.

To settle the controversy, Friedländer used Gram's staining technique in 1883, even before Gram had published it, to visualize both bacteria. Friedländer found that his bacterium was Gram-negative while Fraenkel's bacterium was Gram-positive. Thus the two bacteria were different ones, settling the matter once and for all.

Friedländer's bacterium later came to be called *Klebsiella pneumoniae* while Fraenkel's bacterium was later called *Streptococcus pneumoniae*, both of which are serious pathogens in their own right.

While in Friedländer's lab, Gram was aware of a stain used by Prof. Paul Ehrlich composed of a mixture of analine, water, and gentian violet. After adding iodine mixed with potassium iodide (known as Lugol's solution) to make Ehrlich's stain bind bacteria better, Gram noticed that the stain sometimes came off some bacteria after washing with an alcohol solution, making these Gram-negative cells difficult to see under the microscope—because there was no stain. Bacteria are actually colorless. So Gram formulated a method to visualize these colorless cells that lost the crystal violet using a so-called counter-stain, called Bismarck brown or vesuvin. The Gram-negative bacteria could be seen under the microscope. This Gram staining procedure was published in 1884.

The Gram staining procedure today commonly consists of the following steps performed on bacteria that are heat-fixed on a microscope slide:

1) Addition of a primary stain, crystal violet—bacteria turn purple or blue.
2) Addition of Gram's iodine—a mordant that fixes the crystal violet to the bacteria—bacteria are purple or black.
3) Decolorization with ethanol—Gram-positive cells keep the crystal violet (remaining purple) while Gram-negative cells lose the stain (turning colorless) upon the alcohol washing.
4) Addition of a counterstain—usually safranin, which turns Gram-negative bacteria red or pink.

On the microscope the Gram-positive bacteria appear blue or purple, and the Gram-negative bacteria appear pink or red.

Incidentally, two additional types of bacteria later came to be known. One type was called acid fast, which retained Gram's stain but was actually neither Gram-positive nor Gram-negative in its cell wall structure. The other type of bacterium lacked cell walls completely. Thus, most bacteria can be categorized into these four distinctive groups: Gram-positive, Gram-negative, acid-fast, and cell wall less.

With respect to the viruses, they are neither bacteriological nor cellular in the way that you and I know about. The viruses merely consist of a protein coat that surrounds an internal genome, sometimes harboring a membrane envelope. Thus, the Gram staining procedure is not used to characterize viruses.

5) What other inventions or discoveries was Christian Gram known for?

Dr. Gram had conducted experiments demonstrating how red blood cells became enlarged during a condition known as pernicious anemia, which occurs in patients with a deficiency in vitamin B12 and results in a lowering of the red blood cell count.

In 1890, Dr. Gram went to the University of Copenhagen and became a professor of clinical medicine and pharmacology. During this time he took great interest in teaching, and his course lectures were eventually published into several large volumes, which became popular for teaching clinical medicine by other faculty throughout Denmark.

6) What have I neglected to ask about his life and work?

During Dr. Gram's ensuing years as a clinical faculty, after his great discovery, he formed a thriving medical practice. During this time, 1901 - 1921, he was the chairperson of the so-called Pharmacopoeia Commission and medical director of his department at Frederick's Hospital, until his retirement in 1923.

Dr. Gram was celebrated widely, garnering many accolades and honors. In retirement, he had a keen interest in the history of medicine. In 1938, Dr. Gram died on November 14.

It may be of interest to your readers to consider briefly what subsequent studies showed regarding the biological mechanism that explains why Gram-positive bacteria retain the crystal violet stain while Gram-negative cells lose it.

The first hint emerged in 1929 when Victor Burke and Mildred Winchester Barnes proposed that Gram-negative bacteria had an increased permeability in their cell walls, meaning that the crystal violet stain simply readily exited through the cell wall while such permeability did not occur in Gram-positive cells.

In 1963, this permeability notion was confirmed experimentally when Prof. Milton R. J. Salton definitively showed that alcohol treatment created small pockets in a protein-sugar complex called peptidoglycan that trapped the crystal violet in cell walls of Gram-positive bacteria; while the peptidoglycan of Gram-negative apparently bacteria did not make these pocket traps, thereby allowing the crystal violet to escape.

Chapter 14

THEODOR ESCHERICH

1) Dr. Varela, I know it may seem strange but there seems to be an *E. coli* named after a great scientist named Theodor Escherich. How did this come about?

The *Escherichia coli* bacterium was named as such to honor its discoverer, Dr. Theodor Escherich. He was the very first investigator to isolate the microbe in a pure state, to grow it, and to study it in detail.

In Escherich's time, around the 1880s, and for prior generations, infant mortality had been a chronically serious concern, and the death rates due to infant diarrhea had been particularly alarming, especially for Dr. Escherich, a bacteriologist and pediatrician practicing in a medical clinic in Germany.

As a result, Dr. Escherich had become interested in knowing about the kinds of bacteria that reside in the feces of infants. He had previously learned basic bacteriological methods from Wilhelm Frobenius, who in turn had learned of the techniques from the famous Dr. Koch himself. Thus, knowing how to culture pure versions of bacteria and how to fundamentally characterize his pure bacteria, Dr. Escherich used his newly acquired bacteriological skills to obtain samples of bacteria from the diapers of healthy infants.

Amongst the many types of bacteria found in the infant feces, Dr. Escherich had described a brand new bacterium that he called *Bacterium coli commune,* or *B. coli* for short. Because the new microbe was a common one, Dr. Escherich tacked onto the name the term *commune*. This was the microbe for which he would later be honored by his grateful colleagues by naming the bacterium after him. This was a great honor, indeed.

2) Now, where and when was this famous scientist born?

Theodor Escherich was born on November 29, 1857 in Ansbach, a Bavarian state of Germany. He had dual citizenship in both Germany and Austria.

3) What was his childhood and early years like?

Escherich's father, Ferdinand, was also a pediatrician, as well as a statistician; and Escherich's mother, Maria Sophie Frederike von Stromer, was the daughter of a prominent Bavarian military officer and third wife of Ferdinand. Theodor's father had also been quite concerned with the alarming mortality rates observed among infants and with the inadequate quality of healthcare that had been provided to the poor.

Apparently, these concerns had been handed down to the younger Escherich and consequently had been an important influence in his later years.

Unfortunately, Escherich's mother passed away when Theodor was only 5 years old. It had been reported that due to his behavior problems associated with instigating pranks (it is unclear what the nature of these pranks were), the young Escherich was sent away to a Jesuit boarding school to finish his education. Escherich had subsequently entered the military and served in the Austrian army. After having completed his military term of service Escherich studied medicine, earning his doctorate in 1881, with top grades.

4) For the non-scientists among us, what exactly is *coli*, and is there a difference between coli and *E. coli*?

This is a very good question. When naming new microorganisms, like all other newly discovered living organisms, precise scientific names are assigned to them, for proper classification purposes. These scientific names will then be used to describe the organism by any scientist anywhere in the world when reporting future studies on them in the journals and in books. The first term, *Escherichia*, is known as the genus (plural, genera); the genus is usually a noun (in this case the noun is a person—Dr. Escherich). By scientific convention, the genus is always capitalized and either underlined or italicized, when reported in the scientific literature.

The second term, *coli*, is called a specific epithet; it is usually an adjective. In this case, the term *coli* refers to the source of the *E. coli* bacteria, namely, the colon, and because these bacteria are found in the colons of all living animals, the plural of colon (*coli*) is used. By convention, the specific epithet (*coli*) is not capitalized but is still underlined or written in italics.

Together, the genus and specific epithet, constitute the species of a given organism, or in our discussion, a microorganism (and in this case the species is *Escherichia coli*). Sometimes, in the literature, the species name will be spelled out fully during the very first time it is mentioned in an article, then afterwards the genus is simply abbreviated, e.g., *E. coli*.

5) Why is this particular coli of importance? What impact has this made on science and medicine?

E. coli happens to be an extremely important organism for several main reasons.

First, we know a great deal about this microbe, and this fact, therefore, makes it important as a model for scientific study of basic biology. In fact, I might be so bold as to categorically state that we know more about *E. coli* than about any other organism in the world. Its entire DNA genome sequence has been determined, and we know every single nucleotide base-pair in its genome. We know every gene, and many of its proteins, the amino acids that make up the proteins, and their biological functions; we know many of its other molecules, like sugars, fats, nucleic acids, ions, etc.

Consequently, we know a tremendous amount of knowledge about its physiology, biochemistry and genetics. We certainly don't know everything there is to know about *E. coli*, of course, and we obviously never will know everything about it. *E. coli* is, nonetheless, the best understood organism ever known to humankind, period.

Second, as I mentioned earlier, it is permanently present in the gut of all animals and humans. Thus, we need to know what it's doing there. We know, for instance, that *E. coli* is making our vitamin K, which is quite helpful to us, of course.

It is also very likely that it is protecting us from pathogenic microbes, keeping them in check by effectively competing for limited resources in the gut.

Third, *E. coli* is used by scientists who are molecular biologists as the "laboratory rat" for the cloning of virtually every gene that's studied. Most, if not all, genes, including human genes that are cloned for study, are often put into *E. coli* during the cloning process, in order to closely examine the genes and the proteins that the DNA genes encode.

For instance, this gene cloning into *E. coli* allows scientists to study human genetic diseases in which a defective gene is the causative agent of a given genetic disorder. Many human genetic diseases can be studied by evaluating the physiology and biochemistry of defective genes, their protein products, and those of their normal counterparts, within the *E. coli* bacteria. Thus, *E. coli* is a common and useful tool for basic scientific and medical research. In summary *E. coli* serves to advance scientific knowledge and to improve human health.

Fourth, *E. coli* is similarly used for applied research. For example, when using genetic engineering tools, *E. coli* is employed to help manufacture useful products, such as fuels, plastics, chemicals, medicines, vitamins, hormones, etc.

Fifth, strangely, the presence of *E. coli* anywhere outside of a proper scientific laboratory is an indicator of fecal contamination. For example, if *E. coli* is detected in water or soil, it indicates that these environments have been exposed to human or animal fecal matter. This *E. coli* fecal contamination presents an important issue pertaining to the maintenance of proper sanitation for water, sewage, and the environment.

Lastly, although the *E. coli* bacteria that live in the gastrointestinal tract of all living humans and animals are actually completely harmless microbes, there are, nevertheless, a few *E. coli* variants that are dangerous to humans and other animals. These deviants of the otherwise harmless *E. coli* are pathogenic in that they cause disease (morbidity) and death (mortality) in both humans and animals. They are often transmitted to their victims by the so-called oral-fecal mode of transmission.

One of these pathogenic *E. coli* variants, a strain of the bacterium called O157:H7, caused a serious outbreak of foodborne poisoning from beef that had been contaminated with the strain during the so-called "Jack in the Box" incident in which over 700 victims became terribly ill and 4 children died. Many of the survivors had permanent kidney damage.

There are about half a dozen pathogenic *E. coli* variants. One of these *E. coli* pathogens mimics the effects of cholera, a disease typically caused by another bacterium called *Vibrio cholerae*. Another pathogenic strain of *E. coli* causes a serious condition reminiscent of that caused by the bacterium called *Shigella dysenteriae*, often called dysentery or shigellosis. One newer *E. coli* strain, called NC101, causes colorectal cancer.

Antimicrobial agents, such as antibiotics, are frequently prescribed to patients suffering from the *E. coli* pathogens in order to treat their diseases that they cause.

Unfortunately, many of these pathogenic *E. coli* strains are becoming increasingly resistant to the medicines used to treat them, making efficacious treatment correspondingly difficult to come about.

6) What other great discoveries are attributed to Theodor Escherich?

Dr. Escherich became a noted expert on his "*B. coli*" bacterium. His studies demonstrated much of the early knowledge gained about *E. coli*. For instance, Dr. Escherich was responsible for finding out about how to grow it in the lab and about its bacteriological structure. He used Gram's staining technique to show that *E. coli* is a Gram-negative microbe. Shortly after he discovered the microbe, Dr. Escherich was the first to study its physiology, many aspects of which are still taught in the microbial physiology and microbiology textbooks nowadays.

Dr. Escherich also became an advocate for the improvement and proper healthcare of neonates in order to prevent infant mortality. These issues are still of importance to this day.

7) What have I neglected to ask about this great scientist?

Dr. Escherich's work started what is considered an impressive series of advancements in the sciences. The magnitude of influence by this tiny microbe in science is tremendous. One interesting consequence is the significant number of Nobel laureates who earned their prizes using Dr. Escherich's famous *E. coli* bacterium in their studies. It has been reported that about a dozen Nobel prizes involved the famous *E. coli*.

For instance, one of my favorites is the discovery of the so-called *lac* operon, an *E. coli* genetic locus present on its genome that's regulated by inducers (lactose that's converted to allolactose) and a repressor (a DNA binding protein that prevents gene expression).

The *lac* operon also encodes its regulated structural protein machinery, i.e., a lactose transporter called a permease and an enzyme, called β-galactosidase that starts the breakdown of lactose for energy, to allow *E. coli* to grow. This biological machinery makes it possible for *E. coli* to eat the milk sugar lactose. The Nobel for this operon was awarded in 1965 to Drs. Jacques Monod, François Jacob, and André Lwoff, from the Pasteur Institute, in Paris, France, for their discovery of this gene expression regulation system within the chromosome of *E. coli*. Clearly, there is still much work to be done with *E. coli*, and it is Dr. Escherich who started it all.

Chapter 15

Marie Curie

1) Professor Varela, I suppose if we had 100 scientists in some lecture hall, and asked them to vote on the most famous scientists ever, we would get a vast array of different opinions. But, in terms of one of the most important, pervasively important scientists, whose work has been used literally by millions of physicians, I would have to vote for Marie Curie. What do we know about where she was born and when?

I concur with your assessment of Dr. Marie Sklodowska Curie; not only was she one of the greatest scientists of all time, she is remarkable for her many firsts. She was the first female ever to have earned a doctorate in physics at the Paris-Sorbonne University or anywhere, first scientist to introduce the term radioactivity, first scientist to discover the elements radium and polonium, first female professor at the Sorbonne, first woman to have garnered a Nobel Prize, first person in history to be awarded two Nobel Prizes, each in different areas, one in physics and the other in chemistry, the first woman to become a member of the French Academy of Medicine, and the first (and only) woman thus far to be interred at the famous Pantheon in Paris.

Prof. Marie Curie was born as Marya (nicknamed Manya or Mania) Salomee Sklodowska on the 7th of November in the year 1867, in what is present day Warsaw, Poland, but what was then ruled by Russia.

2) Now her parents, as I understand were both skilled and accomplished in their own right. What kind of environment did she grow up in?

Curie's parents, married in 1860, were teachers; her father, Wladyslaw (or Vladislav) Sklodowska, was educated in the areas of mathematics and physics and taught in a Russian gymnasium (secondary) school. Later, after falling into disfavor with his school administrators and consequently receiving several demotions in rank, he started a private boarding school in the Sklodowska household. Curie's mother, Bronislava (or Bronislawa) Boguski (or Boguska), was the director of an esteemed school for girls called Freta Street School. Curie's mother was known to be intelligent and musically inclined. Together, the Sklodowskas had five children, with Marie being the youngest. The Sklodowska household enjoyed intellectual and musical pursuits. The Sklodowska children were taught several languages, and classic books were read out-loud as a family. A particular favorite of Marie's was "*A Tale of Two Cities*." Even the children's games were education-based. All children flourished in school, with Marie recognized early on as being brilliant.

Unfortunately, Marie's eldest sister, Zofia (b. 1862) died in 1876 from the dread typhus, caused by a bacterium known as *Rickettsia*. Furthermore, in 1871 Marie's mother contracted the Great Consumption (tuberculosis), caused by another bacterium known as *Mycobacterium*, and upon the advice of physicians, Bronislava left home for a rest cure, which meant being away from daughter Marie for extended periods of time. In 1878 Marie's mother died from the TB after suffering from the terribly consuming disease for over 6 years. Young Marie was only ten years old at the time. The child Marie, being close to her mother, had been devastated by her mother's death and was sent to live with several of her uncles and became a governess while she was a teenager.

3) Her early educational years- what did she study and where?

The early education of Marie Curie is an epic story of overcoming adversity and in a most triumphant manner. Marie Sklodowska initially attended gymnasium at Freta Street School and then a school directed by Jadwiga Sikorska, who was Polish. In this climate the schoolchildren were taught a Polish education incognito and were taught what the 'proper' responses were to Russian inspectors. Marie was then enrolled in school at Russian Gymnasium No. 3. Apparently, the school was run oppressively in that Polish teachings were banned, and only Russian teachings were allowed. In 1883, at the age of 15, Marie graduated from the school with highest honors.

Next, Marie entered a clandestine school, with thousands of students scattered throughout Poland, in which lectures and teachings were conducted in varying locations in an effort to avoid detection by Russian authorities and, thus, constituting a secret "floating" or "flying" school. The school was referred to as the "Floating University." Marie also worked as a governess during these years, earning money to pay for transport abroad to obtain a higher education, as females were not allowed to attend university in Poland at the time.

In 1891, Marie Sklodowska moved to Paris, France, to attend the Sorbonne University, where she lived frugally and endured severely cold living accommodations during the winters. It has been reported that she fainted once or twice from hunger. It is at the Sorbonne where she first changed her name from Manya to its equivalent Marie on her registration forms. She majored in Physics, and one of her professors was the eminent Henri Poincaré, famous for his work in probabilities, differential equations and topology.

At the Sorbonne, Marie was one of about only 20 females out of over 2000 students total in her college of science, and she was only one of two female students majoring in a scientific discipline. As a student, Marie was extremely hard-working and diligent, even though lectures and mid-term exams were on a volunteer basis, and attendance required a 2-hour daily commute by bus. At the time, final student rankings were made public, and Marie was at the very top of her class after her first year at the Sorbonne.

In 1893, Marie took her degree in Physics—the first female to do so in the school's history, ranking at the very top—first also in her graduating class. In 1894, Marie graduated with her degree in Mathematics near the top of the class, where she ranked 2nd, disappointed that she had not been ranked 1st. It has been reported that these degrees were the equivalent of master's degrees in their fields.

Many years later, in early 1898 Marie Curie began her doctoral research at the University of Paris in the laboratory of Prof. Gabriel Lippmann. Curie studied the newly discovered "Becquerel or uranium" rays. She first had to find adequate laboratory facilities, and then she had to make painstaking measurements of the low-level electric fields. It was arduous work.

She concluded that the rays emitted from the uranium samples were originating from the uranium atoms. She repeated her studies on thorium-containing material. It is here that she came up with the new term she called radioactivity, to denote the energy emitted from radioactive materials. In June of 1903, she received her doctorate in physics, with honors — the first female ever to receive a doctorate in the country of France. Two of the three professors on her graduate committee were themselves later to receive the Nobel, and one of these evaluators is recorded to have stated that Curie's thesis project represented one of the greatest scientific contributions ever made.

4) It is almost as if her entire life led up to the discovery of X-rays, and the capacity to see inside the human body and skeleton. How did it start and how did it come about?

The X-rays themselves had been discovered by Wilhelm Conrad Röntgen in 1896, earning the Nobel Prize for the work in 1901. This led to Dr. Curie's interest in X-rays and made her consider working on them for her Ph.D. thesis project. However, she chose instead to study the Becquerel rays, which in turn led her to discover the radium, an extremely important finding in and of itself.

Another important connection between the X-rays and Curie involved her work during the Great War to develop mobile X-ray units for direct battlefield examinations of wounds for bullets and shrapnel. Curie herself designed and built the mobile units—she put together the various components, such as the actual vehicles, an electric generator, batteries, an X-ray tube, a collapsible gurney for patients, plates, a screen, light shielding curtains, and radium gas ampules. These battlefield-ready mobile X-ray units, called "*Les Petites Curie*" were built, and Curie and her eldest daughter, Iréne, then a teenager, actually drove the units, deployed them, and used them extensively throughout the duration of the war. The Curie mother and daughter duo also trained others to perform the same activities with the X-ray mobile units. Overall, the invention was a significant advancement in military medicine and saved countless lives, and limbs from amputation, during World War I.

5) When exactly was her accomplishment announced and when did it start to revolutionize medicine?

There are two primary Nobel Prize-worthy accomplishments made by Dr. Marie Curie. The first was the discovery of the radiation phenomenon, earning, in 1903, the Nobel Prize in Physics, which Dr. Curie shared with her husband Prof. Pierre Curie (m. July 26, 1895) and with Dr. Antoni Henri Becquerel. The second involved her discoveries of the elements polonium and radium. For these latter efforts, Curie received her second Nobel Prize, this time in the area of Chemistry, in 1911. She was the sole recipient; her husband had died tragically in 1906 after he had been run over by a horse-drawn wagon.

In terms of advancements in medicine, radiation has been used in a variety of ways. The use of radium and other radioactive isotopes have been used as so-called last resort therapies to treat cancers. Radiation has also been used to detect important biomolecules, tracing them through cells, tissues, or through the body, for instance. Certain molecules can be labeled with radioactive atoms in them and then used to learn new properties of important physiological processes, such as transport of neurotransmitters, sugars, and of other metabolites across the cellular membrane.

Many molecular biological and biotechnological techniques have been developed based on the use of radioactivity. For example, the so-called radioimmunoassay (RIA), has been used to measure antigens and antibody interactions. The Nobel Prize was awarded to Dr. Rosalyn Yalow in 1977 for her development of the RIA method in order to study hormone peptides. Incidentally, Dr. Yalow had been inspired to study atomic physics after reading Eve Curie's biography of her mother Marie Curie.

6) How much recognition was she given during her lifetime?

Dr. Marie Curie did not receive much recognition during her formative years, even when she was rightly deserving of them through her own efforts. That changed overnight with the Nobel. Later in life, she received more recognition than she apparently cared for, especially after having earned great recognition with the first Nobel Prize. The Curies were both under the watchful eye of the public and admirers. They were constantly pressed for interviews or to make speeches, etc.

Often when Curie was travelling, it was for terribly important reasons, e.g., in one incident, she was transporting all of France's supply of radium away from potential capture from the Germans during the Great War, and she remained incognito, even though she was already famous and readily recognized. If she were asked if she was the famous Madame Curie, she would frequently reply that she was not. There were tributes in her honor, fundraisers in her name, and admission to numerous scientific societies that had been primarily for men-only. Quite rightly, she was widely celebrated, with visits to the White House, establishment of Curie Institutes in Poland and in France.

7) What other discoveries was she credited for?

In addition to discovering the polonium and radium elements and theorizing on the concept of radioactive emissions, she was also key to the isolation of radium in a pure state. Regarding radiation, she showed that there was more to the atom than an unbreakable element—smaller particles could emanate from atoms. She also was first to hypothesize that radium could be effective in selectively killing tumorigenic tissue compared to their non-tumorigenic healthy counterparts. She also discovered that the gaseous form of radium, called radon at the time, could be used as an antiseptic to treat infected wounds.

8) Her accomplishments are all that more astounding- since she was a woman, and during that time period, technology to support her was more than minimal. Can you describe her lab conditions and general time frame?

In short, the conditions of her first laboratory were atrocious. Seeking facilities in which to conduct her radium isolation project in the late 1890s, she was provided with what can best be described as a shack. It was actually an old unheated non-air conditioned aviation hangar that had been used for human anatomy dissections but was abandoned because the roof leaked, and the cadavers kept getting contaminated with dust. The pitchblende material from which the radium would be isolated from was brought in by the tons, only to produce minute quantities of radium, many orders of magnitude smaller in quantity than what was initially brought in. The crude pitchblende material had to be heated, cooled, treated with noxious chemicals and re-heated in a seemingly endless process. The resulting purified radium was extremely radioactive, and, thus, it was unknowingly toxic—this was the very first time in history that individuals were working with radioactive materials, and its dangerous effects on humans were all but unknown. It has been widely held that both Pierre and Marie suffered from radiation poisoning.

9) In your mind, and opinion, were X-rays the forerunner of CAT scans and MRI's and should Marie Curie be given at least SOME credit for laying the foundation for these things?

I think that Marie Curie certainly deserves credit to some extent for advancements in X-ray technologies, especially that of mobile X-ray units—the so-called *Les Petites Curie* units. She also deserves a tremendous amount of credit for her work in advancing radiation therapies for cancer treatment. Radiation therapy for the treatment of cancers is a staple in certain serious cases and has been for many years.

10) Her later years- what were they like?

As a famous but widowed single-parent female scientist in her later years and increasingly suffering the ill effects of radiation poisoning, she was nonetheless hardworking and kept a vigorous schedule. She was key in fundraising efforts aimed at furthering scientific research overall. She worked to establish the Curie Institute in Paris, France and the Radium Institute, in Warsaw, Poland, now called the Marie Sklodowska-Curie Institute for Oncology. She supervised doctoral students, many of whom went on to distinguished careers themselves. She collaborated with her daughter Iréne Joliot-Curie (b. September 12, 1897), who would later receive the Nobel in Chemistry in 1935, along with son-in-law (Iréne's husband Frédéric Joliot) for their work on the synthesis of artificial radioactive materials. On the 4th of July, in 1934, Marie Curie died, at the age of 66.

11) What have I neglected to ask?

Apparently, Marie Curie's laboratory notes and other papers were later found to be highly radioactive! Despite later decontamination procedures, her notebook materials are still considered somewhat hazardous to handle. There have even been contentions that Curie's home cookbooks are radioactive. Apparently, scholars wishing to study the documents must sign certain waivers in order to do so. Curie's personal documents are stored under heavy shielding in order to provide protection from the radioactive papers.

As I mentioned briefly above, Marie Curie had another daughter, Eve Curie, born on the 6th of December in 1904. Although Eve did not become a scientist, she was exceptional in her own right. Eve Curie and her husband Henry Labouisse, Jr., received the Nobel Peace Prize in 1965 for their efforts pertaining to UNICEF (United Nations Children's Fund). Eve Curie also published a widely popular and inspiring biography of her mother.

In 1943, a popular film was released by MGM Studios titled *Madame Curie*, staring Greer Garson as Marie and Walter Pidgeon, as Pierre. The film received numerous Academy Award nominations.

Chapter 16

EDUARD BUCHNER

1) A winner of the Nobel Prize was both a chemist and a zymologist. For the uninitiated- what exactly is or was a zymologist and why is this important?

The term zymologist is an historical term originally meant to convey an individual who studies or practices the art and science of fermentation. Zymologists apply their knowledge of fermentative processes to produce foods, like bread or sauerkraut, and drinks, like wine or beer. The term zymology arises from *Greek* meaning 'to do fermentations.' The term enzyme refers to the fact that yeast microbes were first discovered to conduct the fermentation process. Many other microbes undergo the fermentative process, as well, like certain bacteria.

Historians of science maintain that Prof. Louis Pasteur was the first chemist to become a zymologist, as he was the first investigator to provide evidence that yeast microorganisms performed fermentative chemistry. Today, the term zymologist has morphed into the modern version called enzymologist.

Within the enzymology field of the biochemistry discipline, the investigator studies the chemistry of proteins known as enzymes, which catalyze biochemical reactions by converting a substrate into a brand new product, but with the enzyme itself remaining essentially unchanged during the living biochemical process.

Dr. Eduard Buchner was the first investigator to demonstrate that fermentative chemistry could be performed outside of a living yeast cell by using a yeast protein component, which he knew to be an enzyme activity that he had removed from yeast cells and which he had moved into a test-tube. The Brewer's or Baker's yeast cells are known today as a member of the microbial species *Saccharomyces cerevisiae*. In his 1897 publication, Buchner called the enzyme zymase, which had been studied by others, including his brother, Hans Buchner.

Eduard Buchner then demonstrated the presence of a biochemical process called fermentation, which could actually be carried out with the extracted zymase enzyme. The fascinating breakthrough by Buchner had been that the zymase fermentation activity happened without its living host yeast cell to help it out!

For this historical first-time demonstration of the so-called 'cell-free fermentation,' Buchner has been noted as the first biochemist ever to have received the Nobel Prize, which was bestowed to him in the field of Chemistry, in 1907. Buchner has since been dubbed as one of the founding fathers of modern biochemistry.

2) Eduard Buchner's name is closely linked with the concept of fermentation. Now, what exactly is fermentation, and why is it important in the big scheme of things?

Buchner experimentally demonstrated a yeast-derived cell-free fermentation process. There are several meanings for the term fermentation, depending on who is interested in it.

At its most basic metabolic level, fermentation involves glycolysis and any one of a number of fermentative pathways, producing specific bio-molecular end-products. The fermentation pathway is usually named after the particular end-product formed during the pathway; for example, in beer fermentation the end-product is the beer that's bio-synthesized when the fermentative process is complete.

To those in the food and beverage industries, fermentation refers to the process that's used to make food and drink, like, for instance, cheeses or alcoholic beverages, for consumption. On the other hand, fermentation also refers to the spoilage of food, drink, or other products, such that the spoiled items are then deemed unacceptable for consumption or productive usage. In the food, drink and product industries, extensive efforts are often geared towards the prevention or minimization of such spoilage. To an industrial chemist, fermentation may be considered as describing a large-scale microbe-based process that occurs in the absence of oxygen in the air.

To a microbiologist, the term fermentation is frequently used to describe a microbial metabolic process in which energy is released under anaerobic conditions. To the biochemist, fermentation involves a metabolic process that releases energy in the form of ATP from sugars and other organic substances and which occurs without the need for the Krebs cycle or a respiratory-based electron-transport chain to bring electrons to oxygen, but instead uses organic molecules, such as fumarate, succinate or pyruvate, as final acceptors of electrons.

The process of fermentation is commonly used to make important and useful end-products, such as ethanol for fuel, or beer and wine for consumption. Fermentation pathways can produce acidic molecules, which are present in commodities like, for example, vinegar, cheeses, yogurt, rye bread, sauerkraut, and sausage.

Propionic acid and carbon dioxide molecules that are produced during fermentation biochemistry provide the flavor and produce the holes in Swiss cheese. Fermentation has been employed also to produce glycerol and acetone compounds in the pharmaceutical and chemical industries, respectively. Fermentation is routinely used to produce citric acid for drink and food flavorings. Methane produced by fermentation is frequently used as a fuel. Fermentation of the sugar called sorbose is used to produce large-scale quantities of vitamin C for nutrient supplementation.

3) Buchner provided evidence of cell-free fermentation. Why was this important at the time? Is it still important?

In 1897, Buchner grew up his yeast cells in culture and then broke them open, releasing their internal contents and making a so-called yeast juice extract. His yeast juice contents were free of the living cells that had contained the microbial internal cellular contents, but were without their biological membranes that had surrounded the yeast cells—the cell-free system. To his broken yeast-cell juice extract, Buchner added cane sugar molecules called sucrose to the inner biological machinery of the yeast juice, minus the intact living yeast cell, and found that the yeast workings within the juice extract could readily convert the sucrose sugar into ethanol and carbon dioxide. Buchner's method worked well with the addition of other sugars, like glucose, fructose and mannose, each of which showed good ethanol and carbon dioxide production.

Incidentally, when Buchner added the sugars lactose or maltose, he saw no fermentation, consistent with what he and others had seen with intact yeast, as we now know that the *S. cerevisiae* Brewer's yeast is not capable of utilizing these sugars, unless genetically engineered to do so.

During Buchner's studies of fermentation with the common table sugar sucrose, he discovered that an enzyme was the key to the biochemistry of sucrose fermentation. He attributed this fermentative activity to an enzyme called 'zymase,' which we now know to be a mixture of distinctive enzymes, all of which partake in the fermentative breakdown of sucrose. The modern equivalent of the zymase activities include enzymes like sucrase, or invertase, which break down sucrose into the individual sugars glucose and fructose.

Although its enzymatic activity had already been known for many years and mentioned in the literature by Moritz Traube and Carl von Nageli, other sources report that Buchner's brother, Hans Buchner, had had a role in studying zymase, as well. Eduard Buchner had attributed the sucrose fermentation biochemistry that he had observed to his brother's zymase enzyme.

There is a disease in humans called sucrose malabsorption, or sucrose intolerance, in which individuals cannot metabolize the sugar because the sucrase enzyme is deficient, or lacking in its activity, which is due to an inheritable genetic disorder. The patients will usually have to either avoid consuming foods and beverages containing the sugar or partake in enzyme replacement therapy, taking medication or bio-engineered food or drink containing the enzyme.

Buchner's cell-free fermentation results were important because, at the time, many investigators were still convinced of a so-called 'vital force' which was thought to be a naturally-occurring hypothetical amorphous 'essence' that brought about life to non-living molecules.

Prior to the famous experiments of Buchner, it was widely thought that the fermentative biochemistry that had been conducted by intact living yeast required the elusive 'vitalistic force.' With Buchner's demonstration of the ethanol fermentation occurring without the need for living yeast and its associated vitalism, his work paved the way for the end of vitalism and allowed the entry into the new era of experimental biochemistry.

Buchner's work put an end to the so-called mysticism associated with the vital force of life, which, for many centuries, had heretofore been thought to be required for bringing non-living molecules to life. This was a profound and fundamental change in our way of thinking about what chemical matter constitutes life.

Buchner's work also meant that if enzymes that had been taken out of yeast cells performed the complex chemistry involved in producing alcohol, then perhaps many other complex biochemical workings in living cells used enzymes, too, independent of their organisms of origin. With Buchner having ushering in and establishing the novel field of experimental metabolic biochemistry, many other investigators set about to study the fermentation systems by isolating each of the enzymes involved and in elucidating each of the biochemical steps in the various fermentation pathways.

As a result, investigators Otto Meyerhof and Gustav Embden studied the first seemingly universal steps involved in undergoing fermentation, namely, those biochemical reactions of glycolysis, in muscle tissue. They successfully elucidated the entire pathway, now known as the Embden-Meyerhof glycolytic pathway. An offshoot of the Embden-Meyerhof pathway was discovered by bringing onboard another investigator, Jakub Parnas, and together the trio elucidated a glycolytic pathway found in both bacteria and many eukaryotes. The so-called Embden-Meyerhof-Parnas glycolytic pathway represented some of the most primitive biochemical reactions in evolutionary history, probably originating with our common ancestors and providing a basis for gluconeogenesis reactions. Additionally, Otto Warburg and Has von Euler-Chelpin elucidated the corresponding glycolysis pathway in Buchner's famous yeast juice extract, confirming that zymase was in fact a system of distinctive enzymes.

The cell-free fermentation system of Buchner is still relevant today as modern biochemists and molecular biologists still use the approach to study metabolism of living organisms. In this same way, these scientific investigators isolate, identify, and characterize each of the enzymes involved in the many fermentative pathways associated with microbes and other organisms. Buchner's work led the way towards the elucidations of many other biochemical pathways found in intermediary metabolism, such as those biochemical reactions involved in the metabolism of the fatty acids, nucleic acids, amino acids, and the many types of sugars.

4) What modern implications does it have?

One modern implication presently has to do with the recent emergence of the new field called metabolomics. The new field deals with the metabolic signature that living cells leave behind as they conduct their biochemical behaviors. The new investigative field aims to identify and catalog each of the small biological molecules that constitute the metabolites in, for example, a given cell, tissue, or an organism. Therefore, if one knows the complete list of the metabolites of a cell, or a tissue or even of an entire living organism, that metabolite list is referred to as its metabolome.

Buchner's work demystified vitalism, thus, making it possible to facilitate the study of modern bioenergetics, which is the field devoted to energy generation in living organisms, such as the biological energy that is derived from fermentation or from respiration. Electrons that are taken from foodstuffs are sent to various respiratory chains, creating an energy mode called the proton gradient (or in some cases a sodium gradient) of energy that is, in turn, used to conduct phosphorylation of ADP to produce ATP, another form of biological energy.

Another modern and relevant implication of Buchner's work has to do with the biomedical health sciences. One such example is that related to the study of cancer. The disease called cancer is actually a constellation of many different tumor-based diseases, each depending upon the affected cell or tissue. In tumors, certain cells will grow in an uncontrolled manner, possibly being detrimental to the normal physiological functioning of the cells, tissues or organs, leading potentially to morbidity and mortality. The cellular machinery that is involved in the biochemistry that accompanies cancer cell growth may be suitable targets for modulation. This has been a long-standing and important field of study in the biomedical sciences.

Buchner's work has modern implications in the study of molecular biology, as well. For many years, the molecular biologists have used their molecular tools to learn more about the biology and chemistry of cells. Furthermore, in order to learn new mechanisms in the field of molecular biology, each of the molecular players involved in molecular and cellular biological processes have to be purified to homogeneity, in order to fully understand their molecular mechanisms of operation. Biochemists will be needed in order to purify these molecular players.

5) Like D'Artagnan of the Three Musketeers, Buchner died a somewhat heroic death. What do we know about this chemist who died too soon?

While an artillery officer in the Bavarian army, Buchner served as a Captain in charge of a unit tasked with protecting an ammunition supply depot in 1914. Capt. Buchner served in military campaigns in Northern France and on the Eastern European front, earning the Iron Cross medal and a promotion, in 1916. After his term of service was completed, he reenlisted. On the 13th day of August in 1917 during the Great War, Major Eduard Buchner was killed in action from a shrapnel wound in Romania at 57 years of age.

Buchner had been born in the year 1860 on the 20th of May in Munich, Germany, to parents Ernst Buchner (father) and Friederike Martin Buchner (mother). Young Buchner's father died when Eduard was only 12 years old. Buchner attended grammar school in Munich and then moved to the Munich Polytechnic studying chemistry in the laboratory of the famous Prof. Emil Erlenmeyer—any student of chemistry or biochemistry laboratory courses will no doubt be familiar with the 'Erlenmeyer flask,' named after Buchner's mentor. In graduate school, Buchner had studied under the then famous chemist Prof. Adolf von Baeyer and took his doctorate in chemistry in 1888.

6) What have I neglected to ask?

Some individuals are often under the mistaken impression that the so-called 'Büchner funnel,' which has been used by many an organic chemistry laboratory student, was named after Eduard Buchner. However, the fact of the matter is that the famous chemistry funnel, and the equally famous Büchner flask, are named after the industrial chemist Ernst Büchner, who invented the glassware vessels, patented in 1888.

Eduard Buchner's friend, colleague, and biographer, Carl Harries, has conveyed a somewhat amusing anecdote about Buchner. The story was told by Harries that in Tübingen, on the night before Buchner's ceremonial wedding to his fiancée and future wife, Lotte Shahl, a supremely blissful, but somewhat inebriated 40-year old Buchner almost got arrested for singing rather loudly while walking down the street on his way home from the spirited celebration.

Lotte being the daughter of a prominent professor of mathematics and granddaughter of a prominent philosopher (Friedrich Trendelenburg) and Eduard having been the son of a distinguished forensic pathologist, the putative arrest of Eduard almost certainly would have been a scandal of enormous proportions. Historians may not have been so sure about what might have transpired with 24-year old Lotte, who no doubt would have been rather put out and leery about having to bail her bridegroom out of jail on her wedding day. Luckily for the Buchners, Eduard stayed out of jail, and the wedding was held the next day without any other known hitches.

Chapter 17

ALEXANDRE YERSIN

1) Dr. Varela, today, I would like to ask you about Alexandre Yersin, who apparently was well known for his studies of the plague bacterium that was interestingly enough, named after him. What do we know about his birth and early life?

In 1863, Alexandre Emile John Yersin, known as the discoverer of the plague-causing bacterium *Yersinia pestis*, was born on September 22 in the municipality of Aubonne, located in eastern Switzerland. Just a few weeks prior to his birth, Yersin's father, a natural science educator, died suddenly. Thus, Yersin's mother and his two siblings (Franck and Emilie) moved to Morges, Switzerland, where as a single parent she founded a school for girls. As a child it is reported that Yersin had an interest in the natural sciences, which was the study of the biological, physical, geological, and chemical world. Yersin went to high school in Lausanne, Switzerland. He then began his studies in medicine while in Lausanne but then moved to Marburg, Germany, and later Paris, France, where he took his medical degree in 1888.

Dr. Yersin then took a post in the laboratory of the famous Dr. Prof. Louis Pasteur at the Ecole Normale, having been invited by Dr. Emile Roux to be his (Roux's) assistant in Pasteur's laboratory. While there, Dr. Yersin took part in the discovery of Pasteur's famous rabies vaccine. In 1889, Yersin became a citizen of France, apparently in order to be allowed to treat patients there. Yersin's parents were French, and this may have had an influence in his decision, as well. During that same year, Dr. Yersin spent a few months working in the laboratory of the famous Dr. Robert Koch, in Berlin, Germany, acquiring training as a bacteriologist. Next, Yersin moved back to Paris to work and collaborate with Roux at the newly established Institute of Pasteur, where they studied diphtheria caused by the bacterium called *Corynebacterium diphtheriae* and discovered the diphtheria toxin.

2) I suspect everyone, at some time or another has heard about the "Black Death" and various forms of the plague. In the old days and Middle Ages, obviously sanitation was not the same as today. But what were some of the contributing factors to the "plague bacterium"?

The plague disease in humans is caused by the *Yersinia pestis* bacterium, a bacillus-shaped Gram-negative microbe sharing the same bacterial family as the *Escherichia coli*, *Salmonella enterica*, and the *Shigella dysenteriae* bacteria. The plague is considered to be a so-called zoonotic disease, because it involves transmission to humans from a variety of animals. Fleas, for example, then carry the bacteria from the animals to people.

There are three kinds of the plague. The first is the most common and is called the bubonic plague, characterized by the so-called bubo lesion (the medical term for bubo is adenitis) on, for example, the arms and neck of its patients or in places where lymph nodes are found. The bacterium is transmitted to humans by fleas and other insects having emerged from rodents and other small animals. The pneumonic plague is a severe lung infection also transmitted by fleas and rodents but also by person-to-person contact and via the airborne mode of transmission. The aggressively lethal septicemic plague, where the bacteria actively multiply in the patient's blood, occasionally arises from first acquiring the bubonic or pneumonic plague forms.

In our history, there have been three so-called great plague pandemics, i.e., world-wide presence of the disease, all pandemics of which are well-known and documented. The first great pandemic is referred to as the urban or the Justinian plague, having started in Egypt in 541-545 AD, during the reign of the Byzantine emperor, Justinian I, and lasting well into the 6th and 8th centuries, killing, by some accounts, an estimated 100 million victims in its wake. In fact, Justinian I himself is thought to have suffered from the plague, but he survived the disease. It is further believed, however, that the Justinian plague pandemic contributed to the eventual fall of the Roman Empire.

The second plague pandemic, called the medieval plague or "Black Death" by later historians and the "Great Mortality" by its victims who perished from it, started approximately in 1320-1353 and lasting well into the 18th century with multiple epidemics throughout Europe and Asia, killing an estimated 20 to 200 million people, decimating, in particular, the European population and profoundly affecting its societal stability.

The term "Black Death" has an uncertain origin. Some say the term arose because of the black color of the dead tissue apparent in the septicemic patients, but this explanation is suspect because the septicemic plague was relatively rare. Another explanation holds that the Black Death term arose because of an error in translation.

The story starts with the Swedes in 1555 and the Danes in 1605, and later by historian Johannes Isaacus Pontanus in 1631, all of whom were apparently under the misapprehension that the Latin terms used to describe early plague epidemics by Seneca and de Corbeil (*atra mors*), or by Simon de Covino (*mors nigra*), meant "Black Death" or

"Big Death" when these early writers probably more likely meant to say "Great Mortality" after the shockingly substantial numbers of deaths associated with the plague. In any case, the origin of the term "Black Death" remains a mystery.

The third, or modern plague pandemic, is said to have started in 1894 in Hong Kong, lasting well into the early 20th century, and killing an estimated 200 million people. Some believe this pandemic is present today.

It is this pandemic, however, from which Yersin is said to have isolated the bacterium known as the causative agent of the plague disease.

Altogether, it is widely believed that the plague has wiped out approximately one-fourth to one-third of the world's human population, thus permanently altering the course of history. The significance of this *Y. pestis* microbe, therefore, cannot be overstated.

3) How did he come to discover this bacterium that was ultimately named after him?

The story of how Dr. Yersin discovered the plague-causing bacterium begins when he travelled to Hong Kong in June of 1894, shortly after the start of the modern or third great plague pandemic. Yersin's purpose in making the trip was to investigate the emerging outbreak there of the bubonic plague, which by the time Yersin arrived had already killed an estimated 60,000 people there.

Yersin was not immediately welcomed in Hong Kong, however, as local officials had already appointed the well-respected and very famous Japanese physician and bacteriologist Dr. Shibasaburo Kitasato, who had arrived in Hong Kong three days earlier, to lead the investigation. Dr. Kitasato had worked in the eminent Robert Koch's laboratory and had discovered the tetanus-causing bacterium, and he had worked in Emil von Behring's laboratory, developing potential antisera therapies for the tetanus, anthrax and diphtheria—all serious diseases.

Kitasato had even been nominated for the very first Nobel in the category of physiology or medicine. Consequently, Yersin, prevented from conducting autopsies, had trouble even obtaining plague tissue samples, reportedly going so far as to bribe handlers of dead bodies of the plague victims to "appropriate" specimens. Eventually, perhaps reluctantly, Yersin was allowed to proceed with his investigation, but only after repeated appeals to the local Hong Kong officials.

Once fully accepted, Yersin went to work with his investigation of the outbreak.

First, he acquired samples, properly this time, of the buboes tissues from the autopsied corpses of the Hong Kong plague victims. Next, he found bacteria within the buboes tissues while studying them under the microscope. They were bacillus-shaped (i.e., rod or stick-like) in his microscope. Yersin then isolated the bacilli bacteria by culturing them on a special peptone agar culture medium in an incubator. Next, using his purified bacterial cultures, he inoculated test animals, guinea pigs, mice and rats, with his purified bacteria. The inoculated laboratory animals died. Then Yersin examined the tissues and organs of the dead animals and found the same bacillus-shaped bacteria in these dead animals, using his microscope.

He quickly returned to Paris and shared his findings, presenting them first at a scientific conference (a congress) and then publishing them in the journal "*Annals of the Institute Pasteur*," in 1894. It is reported that Yersin even personally informed both Drs. Roux and Pasteur of his discovery.

4) What were the initial reactions?

In short, the reactions were mixed. In certain circles, Yersin's plague bacillus was immediately accepted as the causative agent of the second great pandemic. On the other hand, Dr. Kitasato had also essentially performed the same type of investigation as Yersin, identifying the plague bacterium from the blood of plague victims and animals. In fact, Kitasato had even published his work in the prestigious journal *Lancet*, prior to Yersin's publication. All told, it is becoming increasingly clear that both Yersin and Kitasato deserve credit as co-discoverers of the plague bacterium, each investigator having worked independently of the other.

It is not entirely clear why Yersin, and not Kitasato, received the bulk of the credit for the historical discovery. The microbe was, after all, named to honor only Yersin. One explanation that has been put forth is perhaps that Kitasato's preparation was not completely pure and contaminated by other bacteria, in particular, with the so-called *Streptococcus*.

Another idea is that, in his paper, Kitasato had stated in Japanese that his bacilli, while appearing like *Streptococcus*, were not to be mistaken as such, as the bacteria were bacillus in shape, but perhaps that that proper phrasing was missed in the subsequent translation. The literature today frequently lists both Yersin and Kitasato as co-discoverers of the plague microbe.

At the time of his publication, the plague bacteria were named by Yersin as *Pasteurella pestis*, but other names had been given for these same bacteria, such as *Bacterium pestis*, *Bacillus pestis*, and *Pestisella pestis*. The term "pestis" is Latin for plague. The scientific name we have come to know today, *Yersinia pestis*, was established in the mid-20th century, again mainly as a tribute to Yersin.

After his historical discovery, Yersin focused his collaborative research efforts with Drs. Albert Calmette and Emile Roux on developing a plague antiserum, which unfortunately had poor efficacy. Yersin moved to Vietnam where he established a school of medicine. Yersin died in 1943 on the first of March, at 79 years of age.

Kitasato, on the other hand, while losing out on the Nobel in 1901 to his collaborator, von Behring, for their work on diphtheria, nonetheless made other significant discoveries.

For example, with his student, Shiga Kiyoshi, they isolated the causative agent of dysentery, the bacterium known as *Shigella dysenteriae*. Seventy eight years old, Kitasato died in Tokyo, Japan, on the 13th of June in 1931.

Neither Yersin nor Kitasato received the Nobel for their discoveries.

For many years, there was a controversial issue about whether the *Y. pestis* was the causative agent of the first two plague pandemics, the Justinian and the medieval plagues, as well as the third pandemic, from which Yersin and Kitasato had demonstrated. In the first two instances, corpses from both of these pandemics had been exhumed from their graves, and their teeth pulp examined for the presence of the *Y. pestis* contagion.

After all of these years, only the teeth pulp is thought to harbor any remaining plague bacteria. The *Y. pestis* bacteria were indeed found in the samples tested. Thus, because *Y. pestis* has been definitively shown to be responsible for the emergence of all three pandemics, it demonstrates that this one pathogenic bacterial species alone has killed more human beings in history than any other contagion known to humankind.

5) What impact does this discovery have on us today? Are there any variants of these bacteria still in existence, and are they resistant to antibiotics?

Oddly enough, the plague remains only a relatively minor problem today, with approximately 10 cases per year in the U.S. Beginning in 2011, the DNA genomes of the *Y. pestis* bacteria from each of the pandemics were pieced together and later completely sequenced and compared to the genomes of modern *Y. pestis*. Surprisingly, the ancient genomes did not differ widely with its modern genomic counterparts. In fact, only about 100 nucleotide differences were observed between the ancient (deadly) and modern (not so deadly) *Y. pestis* genomes.

The question, thus, arose as to why the ancient form of the bacterium was so deadly in ancient times while its modern and mostly genetically identical descendant is, relatively speaking, virtually harmless today. The immediate obvious answers are that the fleas, rodents and other so-called vectors of disease transmission are now sufficiently controlled to block the contagion today; another idea was that perhaps all of the people who succumbed to the plague and died could no longer pass along their susceptible genes into the next generations while those who survived were able to propagate their immune genes into the succeeding generations, and now modern human beings are consequently and simply immune to the plague.

A third notion is that improvements in medicine, such as antibiotics, and vaccines, plus sanitation, and personal hygiene practices, were all effective enough to countermand the modern plague.

Evidence for a fourth explanation, however, was published in 2015 to account for the vast differences in virulence between the ancient and modern plague bacteria. In the study, investigators closely examined the DNA sequences of the ancient versus the modern genomes of the two *Y. pestis* bacterial forms and found two sequence variations between the ancient deadly and the modern harmless bacteria.

In the first genetic variation, the ancient plague bacterium acquired a new plasmid containing a gene, called *pla*, which encodes a so-called plasminogen activator protein, Pla, a protease enzyme that breaks-up another inactive protein, called plasminogen, thus converting it into an active form of the protein, called serine protease plasmin. This active plasmin then allows the ancient bacteria to escape from clots that are originally meant to contain the microbes. Now the escaped bacteria move on to infect the lymph nodes and grow within them, swelling up and forming the characteristic buboes.

The second variation of these ancient *pla* gene-containing microbes had a new mutation at a single-point within the *pla* gene that made the resulting protease work even more efficiently than before in order to greatly enhance the virulence of the ancient plague strain. The new mutation caused the bacteria to convert from a pneumonic form to a more deadly form that's readily transmitted by coughing or flea bites.

Thus, even though Yersin's plague bacterium has a variety of virulence factors to cause disease, subtle sequence variations in the ancient plague bacterium genome were all that were needed to have a profound enhancement in starting pandemics, in order to further its deadly effects. Somehow, the dangerous ancient bacteria lost or neutralized their deadly genomic variations to pacify their modern counterparts, making the plague a rare infectious disease nowadays.

One potentially alarming impact has to do with the notions of bioterrorism and of the "weaponizing" of the plague bacterium by certain countries. Thus, care, prevention, and bio-defensive measures are often called for in circumstances where individuals working in this area may be at risk.

Regarding your inquiry about antibiotics and resistance to these agents, modern plague is often effectively treated with the antibiotic streptomycin. If streptomycin resistant variants of *Y. pestis* emerge, however, then alternative therapy is invoked with the chloramphenicols, sulfamethoxazole-trimethoprim, and the tetracyclines. In general, many bacteria have developed resistance mechanisms to all of these antimicrobials, but do not appear to be a significant problem for confounding treatment efforts against the modern plague, yet. In some geographical pockets, drug- and multi-drug resistant strains have emerged, but spread of infection through these populations will be prevented with vaccines.

6) What have I neglected to ask?

I think one interesting incident to point out is that, of the many hundreds of millions of deaths which were, in fact, due directly to the *Y. pestis* plague bacterium, it had been mistakenly blamed for a serious outbreak of cholera in London during the 1850s. The rumors began when the excavations had been started for London's new sewer system.

Apparently, during the dig, a large number of human skeletons had been unearthed. Some of the corpses were still in the process of decaying.

One London area in particular was suspect, called the Golden Square district. Prior to its development the same area had been called Craven's Pest Field, and 200 years earlier it had been the site of a mass grave and burial grounds from the multitude of plague deaths that had occurred. The field contained the human remains of about 4000 bodies, buried unceremoniously in mass pits. Craven's field had been left untouched for many years because of the fear of releasing the contagion. Until, of course, urban development encroached, making the area the new Golden Square district, with townhouses. The area was now suffering a severe cholera epidemic in the late 1850s.

Some had speculated that the ghosts of the Great Mortality had had their resting places disturbed by the sewer excavations, or that because of the humiliating way in which they had been buried (aligned improperly and in mass graves) the ghosts had wreaked their vengeance by bestowing the cholera upon the Londoners of the Golden Square district.

Another explanation (which was more reasonable but nonetheless incorrect) was that the sewer construction had inadvertently contaminated the city with plague contagion from the physical disruption of the Black Death burial site. To provide evidence that there was no plague contamination from the sewer project, a map of London had been drawn showing the plague pit, the sewer lines, and the houses of cholera victims. There was no connection between the plague burial pit and the cholera or the sewer lines. There was, however, a connection between the water supply and sewage contamination of the cholera contagion, but that's another story.

Chapter 18

FREDERICK GRIFFITH

1) Bacteria seem to be all around us- and there are good bacteria and bad bacteria. Frederick Griffith seemed to discover something going on with bacteria and DNA. Can you first tell us why some bacteria are good, and others not so good?

This is a very good question. Bacteria are single-celled living microbes which are indeed present around us, virtually everywhere. An entire bacterial organism consists of a single living cell. The bacteria do not have an internal nucleus, a situation that is called prokaryotic. In general, there are two main domains of prokaryotes, those called the archaea, for ancient bacteria (previously designated as archaeobacteria and archaebacteria) and the eubacteria, for true bacteria.

There certainly may be what one may call, as you say, good versus bad bacteria.

Although estimates vary widely between 90 and 99.9%, many microbiology textbooks denote that the vast majority of the bacteria are considered "good" in the sense that they are harmless, or perhaps simply just benign, or maybe even useful in some way. On the other hand, then, between 0.1 and 10% of the known bacteria are thought to be "bad" in the sense that they are, for instance, pathogenic, i.e., disease-causing, or perhaps they could very well be pathogenic, given the right set of circumstances.

Examples of ways in which bacteria may be thought of as "good" are briefly considered here. First, it is widely thought by microbiologists that the archaea were the first living organisms on Earth. This is considered good in that evidence shows these ancient beings conducted activities that made it possible for other life to be possible, as well. For example, the archaea may have "invented" photosynthesis, allowing a good energy making mechanism from light, for later higher plants to use in order to live. Additionally, the ancient bacteria developed enzymes that neutralized the degradative activity of oxygen, called oxidation. In short, because the archaea were the first living beings, they made it possible for other organisms to live, too.

Second, the good bacteria (and other microbes) are necessary for all other life on Earth to continue. One way in which this is possible is that the bacteria recycle many of the elements needed for life, like carbon, oxygen, nitrogen, sulfur, etc. Without these elemental cycles, human life on Earth would be impossible.

Third, good bacteria make substances we may like, desire, or need. For example, bacteria help make foods, like sauerkraut, pickles, cheeses, breads, or chocolate. Bacteria may help produce beverages, like coffee and wine. Bacteria are quite often used in biotechnology to make diagnostic tests for diagnosis of infectious diseases. Bacteria in the human gut may make our vitamin K. Some bacteria produce antibiotics which we can use to treat infectious diseases.

Fourth, bacteria may also constitute part of the overall microbiome on a person or in some environmental niche. As part of the human microbiome, for instance, many species of bacteria keep certain pathogenic bacteria from gaining a foothold and causing disease.

Fifth, some bacteria may live on the roots of plants, helping the plants to acquire important nutrients, like nitrogen, from soil in order to permit the plants to synthesize protein or nucleic acids for plant and crop growth.

Lastly, some bacteria, called probiotics, are considered so beneficial that they are packaged up in foods, like yogurt, or in pill form; these bacteria are meant to provide health benefits in the human gut. Some recent studies have shown that certain gut bacteria affect moods, with certain bacterial types associated with happiness and others with depression, etc.

On the other hand, bacteria that we may consider to be bad are those in which pain and suffering are caused during infectious disease. As a whole, bacteria that are pathogens have been responsible for more death and disease than any other microorganisms. The term 'pathos' means 'to suffer.' Pathogenesis means to make suffering or disease, which is considered any deviation from a healthy state. Infectious disease caused by bacteria may vary from mild to moderate to severe to even lethal.

Pathogenic (bad) bacteria may be transmitted to humans and other animals via a variety of modes. Such transmission modes include foodborne, soil-borne, airborne, vector-borne, person-to-person, blood-borne, oral-fecal routes, etc.

In certain circumstances, bacteria that are normally benign may occasionally take advantage of an opportunity to become pathogenic. These so-called opportunistic bacteria may then grow and predominate, causing pathological consequences in a patient. As one example, harmless bacteria residing on the skin and serving to keep other dangerous bacteria from taking over, might find themselves having become moved to deeper tissue layers or even the blood, from a minor trauma such as a scratch, an insect bite, a needle injection, a surgical incision, etc., and now this previously harmless, helpful bacterium grows unchecked in the blood or internal tissue wreaking havoc and causing disease.

Another way in which so-called bad bacteria mediate their effects is through environmental contamination and pollution. Bacteria play a key role in causing the production of foul-smelling organic compounds that are indicative of polluted environments. Such environments are non-productive and even toxic to human health.

One last example includes bacterial contamination of commercial products, like pharmaceutical items, plastics, blood products, eukaryotic cell cultures, medical and dental equipment, industrial equipment, spacecraft, and commonly used items like cell phones, etc. These types of bacterial contaminations may affect the quality of the products and of the resulting profits, and of commerce, in general.

2) How do bacteria change and grow and mutate and develop over time?

Interestingly, when bacteria grow, they exponentially increase in their numbers rather than get bigger or older. Yet, their generation times, i.e., the time it takes to double their bacterial cell numbers, are quite shorter than, let's say, that of humans. Bacterial growth generations may be as short as 20 minutes. Other bacteria may take a few days to grow, depending on the species or growth conditions. Bacteria often change and mutate when they grow.

Bacteria have developed several ways to change and mutate. Bacteria will change by behaviorally responding to their environment and by mutation. Behaviorally speaking, if bacteria come across a new nutrient, they may turn on cellular machinery that's specialized to acquire it and metabolize it for energy to grow. Sometimes, the nutrient utilizing cellular machinery is already turned on, staying on, especially in cases where there may be a universal nutrient that is available, like the sugar glucose.

A mutation is defined as an alteration in the nucleotide base sequence along a string of double-stranded DNA within, e.g., a genome or a plasmid molecule. The organism harboring the mutated DNA is called a mutant. When bacteria change by mutation, they have developed several ways to do so, some of which are clever but simple.

One simple way in which DNA is mutated is called spontaneous mutation. In this process, the biological machinery that makes DNA just simply makes a mistake when inserting the new nucleotides into the new DNA chains. These types of mutations happen randomly once or twice in every 100,000 or in a million nucleotide incorporations, which is very accurate but not perfect, of course.

Other ways include physical and chemical means of generating mutations in DNA. Physical modes of DNA mutation might include radiation or ultra-violet light. Chemical means of DNA mutagenesis involve changes to the chemistry properties of the DNA that in turn alter the base sequences along DNA.

Importantly, once a bacterium has had its DNA mutated and has, thus, become a mutant, the affected DNA will be transferred to the next generation of bacteria (called vertical DNA or gene transfer) or can be transferred to completely different microbial species (called horizontal DNA or gene transfer).

There are several kinds of these DNA transfer systems. One is called conjugation in which DNA from one generation to the next is propagated into the next generation by mating. The bacterial host receiving the DNA is called a recipient or a conjugant.

Another DNA transfer system is transduction, in which a virus called a bacteriophage (or phage for short) carrying the DNA does the transfer by infecting the bacterium. The infected bacterium in turn propagates the DNA by producing many new viral progeny, each new phage in turn can move on to a new bacterial host and start the infection cycle all over again. The bacterial host getting the new DNA by this method is referred to as a transductant.

Another mode for the transfer of DNA is called transposition. During this process, a piece of DNA called a transposon or "jumping gene" will separate from its location and either move to another location along the genome, to a plasmid, or actually leave the host to find a new host whereupon the transposon DNA will insert itself into the genome or plasmid of another bacterial species; such a recipient bacterium is called a transposant.

The last DNA transfer mechanism considered here involves the one that Dr. Griffith discovered. It's called transformation. In this system, a host cell becomes competent in order to allow the uptake of the foreign DNA, becoming a transformant.

Each of these DNA or gene transfer methods will manifest their new properties to the new host cells that take up the new foreign genetic elements. It is through these systems that bacteria mutate and change. Since the generation times are short bacteria can be altered just as quickly.

3) Frederick Griffith lived in the 1920's (probably long before the electron microscope) How did he do his research and what did he learn?

Dr. Griffith's famous 1928 experiment is routinely included in a fair amount of detail in most major textbooks dealing with Biology, Biochemistry, Cell Biology, Genetics, Microbiology, Microbial Physiology, and Molecular Biology. Griffith studied two versions of a bacterium that investigators at the time had called *Pneumococcus* but is now known as *Streptococcus pneumoniae*. The two bacterial versions were distinct variants of each other.

The first version was a non-virulent and non-lethal kind of the bacterium which he called Type II or the R form. The R form bacteria were called as such because they formed rough-looking colonies on culture media in Petri plates. The R form was harmless because it lacked a specialized covering called a capsule—it was non-encapsulated, and macrophages and phagocytes could readily protect the body from the R forms of the bacteria; thus the R type was non-lethal.

The second version Griffith called type III or the S form. The S form bacteria were referred to as such because they formed smooth-looking colonies on cultured Petri plates. The S form was harmful because it contained the outer covering capsule. The encapsulated

S form was protected from the macrophages and phagocytes. Thus, the S form bacteria survived in the body, permitting them to grow virtually unchecked throughout the lungs and other organs, causing a severe pneumonia from which the body often could not tolerate, killing the animal or human host. Thus, the S form was lethal.

Griffith injected live mice with the lethal S bacteria, and within days the mice were dead. When autopsied, the dead mice had numerous S bacteria. When Griffith injected living mice with the non-lethal R bacteria, the mice lived, and no R or S bacteria were found in the mice. When Griffith injected into living mice the lethal S bacteria that were first killed by heating, the mice lived—this was not terribly surprising that a dangerous but dead bacterium cannot cause disease.

It is the next step, however, that surprised everyone, including Griffith. He injected into the mice a mixture of live harmless R bacteria with dangerous but dead (heat-killed) S bacteria, and the mice died! Remember, each one alone (i.e., live R and dead S bacteria) did not kill any mice—but together the live R and dead S mixture did kill the mice. When autopsied, the dead mice all had living S bacteria in their blood and organs! But these latter dead mice hadn't been injected with any of the S bacteria, yet there they were in large numbers, living in the dead mice. These data were surprising and unsettling.

How could it possibly be that mice injected with a mixture of two seemingly harmless entities (i.e., live but non-virulent R plus virulent but dead S) be so acutely lethal? Neither entity alone was lethal, but together they were potently lethal. Many explanations were put forth to explain these startling results.

Somehow, something, an "S Substance," as Griffith called it, from the dead S bacteria "transformed" the live R to become lethal S bacteria! What, then, was the nature of this so-called "transforming principle?"

On the face of it, one could simply argue that the live R bacteria stole the capsules from the dead S bacteria. That is to say perhaps the transformation principle consisted of stolen capsules. The problem with this explanation is that Griffith had already established that the S Substance was sensitive to heat but the capsules were not—the capsules were heat resistant. Yet, the S bacteria themselves, and hence the transforming principle, were sensitive to the heat.

Unfortunately, Griffith himself incorrectly reasoned that the transforming principle, the S Substance, was protein; and this protein then made sugar, which then made up part of the capsules—it was known that the capsules contained protein and sugar.

To further confuse the situation, Griffith had then referred to the S Substance as the "S Antigen." In this way, Griffith and others then argued that the R form disassembled the parts (a process called catabolism) from the dead S bacteria (or took their already disassembled S parts) and used the parts to reassemble (a process called anabolism or biosynthesis) the capsules onto themselves to become S bacteria. This explanation was based on the newly emerged field of metabolism—the breaking up of foodstuffs to make new stuff that a bacterium might need for other purposes.

Next, Griffith then argued that perhaps the R bacteria simply mutated to the S form, a process called reversion—the R bacteria reverted back to the S bacteria. He based this explanation on the observation that in hospitals, patients with the pneumonia had several types of the bacteria—so, did the patients get infected with multiple bacterial types or did the bacteria convert to other types? Griffith reasoned that the bacteria may have reverted by mutation from the R to the S forms.

The problem with all of these explanations is that they contradicted the incorrect notion that that bacteria species were unable to switch between sub-types, like R versus S. Therefore, primary opponents of the idea of bacterial transformation, Dr. Oswald Avery and colleagues Drs. Maclyn McCarty and Colin MacLeod, ironically became the first to provide definitive experimental data about the true nature of Griffith's transforming principle: DNA.

Distilled down to the essential and simple basics, in 1944, Avery, McCarty and MacLeod repeated Griffith's shocking experiments, except that this time they included protease (digests protein), RNase (digests RNA) or DNase (digests DNA) in their protocol. They found that the group with DNase failed to transform live R to the S type when live R cells were mixed with heat-killed S cells. With the dead S bacteria DNA being chewed up by the DNase, there was no principle (no DNA) by which to conduct the transformation of the R cells into the S cells. In so doing, Avery and colleagues were the first to identify DNA as the transforming agent. Later, equally famous studies in 1952 performed by Alfred Hershey and Martha Chase using bacteriophages and *Escherichia coli* bacteria to perform transduction, these investigators convincingly confirmed that DNA was indeed Griffith's S Substance, the transformation principle.

4) How do bacteria "transform" (and does heat and cold have any impact on bacteria?)

Bacterial transformation may be defined as the uptake of foreign free DNA that can be incorporated into the bacteria, which in turn can then go on to acquire novel properties conferred by the newly acquired DNA. Bacteria that can acquire DNA are said to be "competent" to do so.

Today, the biological mechanism for bacterial transformation has been worked out for a variety of bacterial species, including that for Griffith's *Streptococcus pneumoniae*. The transformation process for Griffith's bacteria follows below.

First, the bacterium makes a protein called competence factor (CF) that's secreted from the bacterial cell. As the cells grow, the CF proteins accumulate, and, in a process called quorum sensing, CF binds to another protein called ComD that's present on the cell walls of the grown bacteria. Next, the ComD uses ATP to attach a phosphate molecule to itself, and in a process called a phosphorylation cascade, transfers the phosphate to another protein called ComE, and then on to SigH. The phosphate-SigH complex then turns on a set of genes that code for a large molecule called the "transformasome complex." Once a bacterium makes this transformasome, the cell is said to be competent to take up foreign DNA. The transformasome then binds DNA that's nearby the cells and brings it to the inside of the cell, where the DNA can then be expressed, altering the transformed bacterium.

Regarding your question about heat and cold, investigators can instigate transformation artificially in the laboratory using calcium chloride, cold, and heat, in order to have bacteria take up DNA, for cloning or other purposes. Investigators grow bacteria, add calcium chloride, and place on ice, in ice buckets. This artificially makes the cells competent. Next, the investigators add DNA. The DNA enters some of the cells. Next, the cells with DNA are briefly warmed up in a process called heat-shocking; the cells are not killed, but it possibly closes up the cell's walls to lock-in the DNA inside the cell. Then the cells are cultured on Petri plates containing culture media to obtain transformants. Others argue that the calcium chloride and cold merely lets the bacteria bind DNA and that the heat-shocking permits the DNA entry into the cells. Whatever the case is, transformation in the laboratory is an important molecular biology technique.

5) What do we know about the life and times of Frederick Griffith?

Unfortunately, details of Griffith's early life are not entirely clear. Regarding his birthdate, for example, sources vary between 1877, 1879, and 1881. Undoubtedly, he was born in Hale, England. His parents were Joseph and Emily Griffith. Later, Griffith studied medicine at Victoria University in Liverpool, England, and graduated in 1901, with his M.B. degree. He was then a physician and surgeon at the Royal Infirmary in Liverpool. He was awarded an Alexander Fellowship in the area of pathology at a private institute called the Thompson Yates Laboratory. While there, Griffith studied the Great Consumption as a member of the Tuberculosis Commission as a bacteriologist. Acquiring a degree in public health (D.P.H.) at Oxford in 1910, Griffith then became a medical officer in 1911 in London, England, at the Ministry of Health. Tragically, Griffith was killed during an air raid blitz by German bombers on April 17, 1941, during World War II. He was never to know the acclaim bestowed in his honor.

6) In your opinion, why is his work so important?

I think that Griffith's discovery of bacterial transformation led ultimately and directly to the elucidation of DNA as the heredity material and later to the structure of DNA by James Watson, Francis Crick, Rosalind Franklin, Raymond Gosling, and Maurice Wilkins. The focus on DNA as being central to life and being the blueprint for living beings then led to the development of the fields of molecular biology and biotechnology, and later to genomics, proteomics, and now bioinformatics. These are all considered important avenues of study.

7) What have I neglected to ask?

Just so it is clear to the reader, the term bacterial transformation as we have discussed here has a completely different meaning when compared to transformation of non-bacterial cells from eukaryotic organisms. As such, transformation in this other case is taken to mean the conversion of a normally dividing cell into a tumorigenic cell, which may be considered either benign or cancerous. Transformation in this particular case involves an uncontrolled cell growth. This is also an extremely active field of study. When one considers the introduction of new DNA into cells that are eukaryotes, such as plant cells, or fungi, or even human cells in culture, the term is often called transfection.

Chapter 19

ALEXANDER FLEMING

1) Dr. Varela, there are certain names that stand out in science, and today, I would like to ask you about one of the great names- Alexander Fleming. What do we know about his early life and work?

If I may be so bold as to say so, I think that Alexander (Alec) Fleming is one of the most famous and recognized scientists in history. At least I believe this is true amongst other biological and biomedical scientists, especially among the microbiologists and immunologists. In our series of amazing scientists considered thus far, Dr. Fleming is the first to have garnered a Nobel Prize for his scientific contribution. A medical bacteriologist, Dr. Fleming was the discoverer of the antibiotic penicillin, which was one of the very first medications to be widely used in order to effectively treat bacterial infections in human beings.

Alexander Fleming was born on the 6th of August in the year 1881 on a farm called Lochfield, which still exists today near the small town of Darvel, located within the council of Ayrshire, in southwestern Scotland. Fleming's parents, Hugh Fleming and Grace Morton, were modest farmers in rural Scotland. His father passed away suddenly from a stroke in 1888, when the young Alec was only 7 years of age. It has been reported that the young Alec was quiet, but spent a large amount of time outdoors where he sharpened his observational skills; and early on he was deemed to be brilliant. Indeed, using his brilliance, Fleming went on to make what has been described as one of the most significant and history-making discoveries of the 20th century, namely, the discovery of the antibiotic penicillin.

Earlier work, however, included Fleming's discovery of lysozyme, in 1922. Lysozyme is an enzyme that is found in tears and saliva. Lysozyme breaks down the cell walls of certain bacteria, producing a kind of natural protection in humans; the enzyme is also found in egg whites.

The story is told that parents would bring their children to Fleming's laboratory at St. Mary's Hospital where lemon juice would be squirted into their eyes, inducing tear production. The procedure has been described as the "ordeal by lemon." The tears were then collected and tested against *Staphylococcus* and *Streptococcus* bacteria grown on Petri dishes in Fleming's laboratory. The lysozyme was effective against these and other Gram-positive bacteria, but not so much against Gram-negative bacteria. The enzyme became very important in studies of cell wall structure of bacteria.

2) His early education- and experience- was I believe in England. What can you tell us about this?

Regarding Fleming's early education, he attended Loudoun Moor School, with only 12 students, Darvel School, where it is reported that he walked 4 miles daily to attend, and the state-funded Kilmarnock Academy, where Fleming had had a scholarship in order to attend. In 1895, Fleming moved from Scotland to London, England to be with his older siblings and to study business at the Regent Street Polytechnic Institute, known today as the University of Westminster. Graduating at age 16, Fleming took a job as a junior clerk with the transatlantic shipping firm, called America Line.

In 1900, Fleming inherited a sizable sum, £250, from his uncle, John Fleming. With encouragement from Fleming's older brother, Thomas, who was already a physician, the 20-year old Alexander used his inheritance to enter the field of medicine; some reports indicate that Fleming also had a scholarship for his medical education. In any case, starting in 1901, Fleming attended Medical School at St. Mary's Hospital of London University in Paddington, a borough of Westminster, in the center of London. In 1906, Fleming passed his medical qualifying exams to earn his Conjoint Board Diploma, and in 1908 he earned his bachelor's degree in medicine (M.B.) and surgery (B.S.), known also as the Bachelor of Medicine, Bachelor of Surgery degree (M.B.B.S.), with honors.

Deciding to pursue a career conducting research rather than practicing medicine, Dr. Fleming entered the research laboratory of the eminent Sir Almroth Edward Wright, whose laboratory was also located at St. Mary's Hospital. Dr. Wright was known for his discovery of a vaccine to fight against the typhoid fever in 1896. In Wright's laboratory, Fleming gained an expertise in the field of bacteriology, discovering lysozyme and becoming a lifelong bacteriologist. Fleming was to remain at St. Mary's for the entirety of his career.

3) Penicillin - I daresay this word is almost universally known for its potent properties and many uses. Can you give us a few examples?

You are correct about penicillin, and I am glad to provide a few examples; penicillin, shortly after it was discovered and mass produced, was considered to be a life-saving miracle medicine because it acutely and potently killed certain disease-causing bacteria.

Along with a handful of other so-called miracle drugs, such as the sulfonamides, the tetracyclines, the chloramphenicols, and the aminoglycosides, like streptomycin, it helped to usher in a new era, the Antibiotic Era, in the 1940s and 1950s. Before this new era began, patients had known only that simple infections could be deadly. Now, patients were astonished to find out that they could receive an injection or even take a pill and live! Prior to the establishment of the new era, such efficacious treatment of infectious disease had been unheard of, and penicillin was one of these new important medicines ushering in the new antimicrobial era with treatable infections.

Often penicillin is used as a platform by chemists to alter its structure to produce new generations of penicillins with structural variations which may have better anti-bacterial activities. When the structure of penicillin was solved, it revealed its active part, called a β (beta)-lactam ring. Penicillin belongs to the group of antimicrobials now called the β-lactams, and some β-lactam derivatives might actually work better against dangerous infectious disease causing bacteria. This is an active area of scientific research.

As you noted, penicillin has other uses. For example, along with other antimicrobial agents, it has been used in molecular biological circles to aid in the cloning of genes. Penicillin use during these efforts guarantees that the cloned genes stay within the bacterial hosts upon transformation, which is a process that puts the new DNA into the bacteria, such as *E. coli*.

You see, the cloning of new DNA involves using small circular pieces of DNA plasmids, called cloning vectors. These plasmid vectors contain, among other genetic elements, a so-called reporter gene, e.g., a penicillin resistance gene, which permits bacteria harboring them to grow in the presence of the penicillin antibiotic on culture media in Petri plates. Thus, bacteria containing these cloning vectors, complete with their cloned DNA and reporters are selected to grow in the laboratory, and those bacteria without the cloned DNA molecules will not be able to grow, because they lack the penicillin resistance gene. In this way, scientific investigators can select those bacteria that harbor the cloned genes for further study.

Gene cloning, in turn, is extremely useful for many other purposes. Cloned genes may be sequenced, either in their bases along the DNA chain or in their deduced amino acid sequences along their protein chains. The cloned genes may be expressed, producing the gene products, usually protein, and the proteins may be more closely examined. Entire genomes have been sequenced, generating a new field genomics. The cloned genes may also be genetically engineered to produce useful products for biotechnology.

Penicillin has also been used in agriculture to treat infections in farm and food animals, to prevent infection, a process called prophylaxis, and in promoting the growth of poultry and cattle. The growth enhancement properties of penicillin and other antibiotics is controversial. Some schools of thought argue that use of penicillin or other antibiotics kills already susceptible bacteria, leaving behind naturally occurring variants that are already intrinsically resistant to the drugs. In these cases, the drug resistant bacteria grow and thus predominate, leaving the antibiotic worthless in its wake.

Some laboratories have demonstrated that the use of one antibiotic, like penicillin, actually selects for bacterial variants that are resistant to multiple antibiotics. Further, investigators have shown that the genes encoding antibiotic resistance and multidrug resistance mechanisms are transferable to other species of bacteria, thus, allowing dissemination of these so-called genetic determinants of resistance through populations and to other regional areas. These phenomena are supported by the studies from many laboratories generated over the course of at least 40 years. Others argue, in contrast, that little or no evidence supports the idea that antimicrobial use results in antimicrobial resistance, especially in agriculture; they argue that the antibiotic use and the antibiotic resistance crisis is overblown.

In any case, it has recently been decided that, as a matter of policy, those in agriculture will voluntarily minimize the use of antibiotics for animal growth promotion. The European agricultural industry has already banned certain antibiotics, and the frequency of antimicrobial resistant bacterial isolates has been reduced in certain cases.

I think it is interesting to note that Fleming himself was among the first, if not the first, to predict that penicillin resistance would emerge soon after its use. He was correct; penicillin resistant *Shigella* bacteria were found for the first time in hospital isolates of bacterial specimens from patients with dysentery during the 1950s, shortly after widespread use of penicillin began in clinical settings.

Lastly, I think it's important to point out something for which penicillin is NOT used. Antibiotics, penicillin included, are not effective against viruses. Instead, antibiotics are useful in killing bacteria, i.e., they are antibacterial agents, not antiviral agents. So, there's no need to use antibiotics for viral infections, like the common cold, or influenza, or yellow fever, because the antibiotics will not work for viral infections.

4) Basically, as I understand it, penicillin is almost a universal antibiotic- just how strong and potent is it, and why is its use so widespread?

In short, penicillin explodes bacteria which are susceptible to the antibiotic. The cellular target of the penicillin antibiotic is the cell wall building machinery of the bacteria. The cell walls may not be built or may be too weak to support its biological activities. You see, a bacterium that is in the process of eating, i.e., metabolizing nutrients to get energy to build stuff it needs for growth, will produce intermediate metabolites and other molecules, all of which build-up inside the microbial cell. The metabolic activity puts pressure on the cell wall of the feeding bacterium. A bacterium with a weak cell wall structure will not be able to keep its metabolites in check, and the cell will explode, i.e., lyse, causing the cell to break up into pieces and die.

Regarding the widespread and universal use of the antibiotic penicillin, I think several factors have come to play important roles. First, I think that the ease of its production and extraction methods, developed early on by Drs. Howard Florey, Ernst Chain, and Norman Heatley, allowed the penicillin medicine to become widely available for use clinically. Second, as mentioned earlier, the discovery that penicillin encouraged the growth of farm animals quickly facilitated a large demand for the drug and in immensely large quantities needed to feed so many animals. Third, I think that because of the therapeutic benefit conferred upon the patients with serious life-threatening infections combined with its relative safety to the patient, made it a prime miracle drug—that is, one that might be considered a so-called "magic bullet." The concept of the magic bullet arose as the idea of a bullet (actually, a medicine) that could go through a person and selectively kill the contagion without harming the patient. The magic bullet term was coined in 1900 by Paul Ehrlich, a Nobel laureate who envisaged the antibody and an arsphenamine called Salvarsan as prime candidates, but neither of which were practical to use clinically, at the time.

Despite potentially serious allergies to penicillin to a certain percentage of people and increasing frequencies of bacteria which are resistant to penicillin and its modern derivatives, the penicillin class of antibacterials remain extensively used antimicrobial agents.

5) As with many famous people, there is always some famous story. Can you retell it for our readers?

There are in fact several famous stories; one is quite well known, and the others are perhaps not so well known. Let's start with the commonly known story, which is a legend in serendipitous scientific discoveries and is even taught to school children around the world.

This first famous story is legendary and starts out in 1929 with the discovery of penicillin resulting from an accident.

Legend has it that Fleming went on a vacation and, being the absent minded professor, forgot to put away his Petri plates that had had *Staphylococcus* bacteria cultures inoculated on them. When Fleming returned to the lab, he found, still sitting on his lab bench, his plates with overgrown bacteria and, worse, with fungal (mold) contamination. Such contaminations had been (and still are) common occurrences ever since Koch's time in the 1880s. Contaminated plates were often simply thrown away with the originally intended experiments having to then be repeated. It was a real nuisance. This one time, however, Fleming had an insight that changed history.

He noticed that his *Staphylococcus* failed to grow around the fungal contamination area on the Petri dish. Fleming reasoned, thus, that the fungus was producing an agent that was then secreted, thus killing the surrounding bacteria. Conducting a sound experimental work up and following through on this hypothesis, he discovered the penicillin, as follows.

First, he cultured the fungus, or "mould" which was identified as *Penicillium notatum*, although some sources also say *P. chrysogenum* or *P. rubrum*, Fleming being unclear about which mold species he actually had in his Nobel Prize winning paper. Next, he took the liquid broth that the *Penicillium* had been grown in and filtered it. Then, he incubated the penicillin-containing extract (from the mold) on various Petri plates containing a variety of test bacteria, to measure the effects on their growth. Many bacteria failed to grow.

Next, he cultured test bacteria in liquid broth, added the penicillin extract, and observed the bacteria disappear—the bacteria had exploded—a process Fleming called bacteriolytic. The results were dramatic. The penicillin was effective against the *Staphylococcus aureus*, *Streptococcus pyogenes*, *S. pneumoniae*, *Corynebacterium diphtheriae*, and *Haemophilus influenzae*, all of which are Gram-positive bacteria and serious pathogens. In the end he showed that his penicillin agent was both bactericidal (bacteria were killed) and bacteriolytic (bacteria were lysed). Interestingly, Fleming's penicillin was much less effective against Gram-negative bacteria. Publishing his findings in 1929, his discovery made its mark on history.

A second famous story regarding penicillin involves scientists' coats and attempts to produce the substance in mass quantities for use. Fleming, being a bacteriologist and not a chemist, failed to purify the penicillin from the mold. It is here where Florey, Chain and Heatley enter the picture. Being chemists, they were able to develop a protocol for large-scale penicillin production.

They had, however, a dangerous problem during these production efforts of theirs. The problem was that being in Oxford during World War II, there was clear and present danger of being bombed by the Germans. The solution, therefore, was to take a boat to the U.S. where the penicillin production protocol could be followed in safety. Legend has it that Florey, Heatley and two others (Arthur Duncan Gardner and Florey's second wife Margaret Jennings) actually had the *Penicillium* mold spores sewed into the linings of their coats, in order to remain incognito during the voyage across the Atlantic to the U.S. and to keep their important work top secret. If anyone were to be killed or caught along the way, the others were then to take up the important work. Incidentally, Florey made a decision to take Heatley with him and leave behind Chain, causing a rift between them (Florey & Chain) that was to last years. Upon safe arrival to the U.S., the scientists unstitched their coats, swabbed the spores onto culture media, grew them up on Petri plates, and implemented their mass production methods to make the penicillin. This group published their findings in 1940.

As I mentioned above, Fleming discovered lysozyme, which later turned out to be a key component in natural immunity. The story is often told that at the time of his discovery, Fleming had a cold and that he actually used mucous drippings from his own nose to test the lysozyme on his bacteria. Other accounts state that Fleming used the nasal secretions from a patient.

Unfortunately, a myth regarding Fleming and penicillin has been perpetuated throughout the years. I say unfortunate because it is such a great story. The myth has two key elements, both of which have been generally and soundly questioned but nevertheless deserve a re-telling, if only perhaps to dispel future recurrences.

The first part of the myth in question was that Fleming's father saved a young Winston Churchill from, depending on the version, drowning in water or in a bog, and that a grateful father of Winston paid for young Alec's education in return. This is considered by biographers to be most certainly untrue, as ostensibly no reports exist about Churchill nearly drowning as a child.

The second key element in the dispelled myth is that in the early 1940s the now grown up Prime Minister Winston Churchill was saved yet again by one of the Flemings, this time from Alexander's penicillin, during the time when Churchill had had a bacterial infection. Now historical records indeed show that Churchill actually did have an infection of the lungs (pneumonia) at the time and that he had been provided an antibacterial agent, but alas, it was not the penicillin. Records show that it was another agent, a sulfonamide called Prontosil, which is a synthetic red dye that had been discovered by a German chemist, Dr. Gerhard Domagk, who in 1939 had received the Nobel for having discovered it.

6) What were some of the other discoveries and accomplishments of this great scientist?

Well, as you know, Fleming was key to discovering the bacteriolytic enzyme called lysozyme. This is a discovery of such great importance that the enzyme is routinely included in modern textbooks about Microbiology, Immunology, and Biochemistry. New fields emerged from the discovery, such as, for instance, protein structure determination and bacterial cell wall biology.

Above, I briefly commented on Paul Ehrlich's Salvarsan as a possible magic bullet (it wasn't). It was Fleming, however, who had conducted the early clinical studies in London of the Salvarsan on patients with syphilis.

An early accomplishment by Fleming was his study of the battle wounds during World War I (The Great War). Fleming had determined that the nature of the bacteria found in infected battle wounds were *Clostridium perfringens*, the causative agent of the horrible and foul-smelling gas gangrene, plus species of the genera *Staphylococcus* and *Streptococcus* bacteria.

Towards this, Fleming tried applying antiseptics to the wounds but found that the agents did not gain access to the tissues deeply enough to be effective and suggested instead that the antiseptics were actually killing useful white blood cells, which were necessary for tissue repair.

This was a controversial finding as Listerism had by this time taken hold as a regular practice for prevention and treatment of wound infections. The issue was further confounded when, after the war, Fleming was unable to definitively identify the principle antiseptic compound that was purported to be lethal to the white blood cells.

7) What have I neglected to ask?

It is chronicled that the chemists Florey, Chain, and Heatley, grew to be somewhat resentful of Fleming. You see, quite often when these chemists and others made a significant breakthrough of their own, it was frequently Fleming, to his bemusement, who was bestowed the bulk of the credit and many of the accolades. It was Fleming, and never the chemists it seemed, who was often called upon by the press to say a few words or to comment on new findings made by others in the field.

It was Fleming's picture alone, and not those of the chemists, who was featured on the cover of *Time* magazine, on May 14, 1944. Fleming received honorary degrees from Harvard and Princeton. Fleming and Florey were knighted in 1944. To make matters worse, however, only three individuals at any given time are allowed to receive the Nobel Prize; thus, Sir Fleming, Sir Florey and Chain all became Nobel laureates in Physiology or Medicine in 1945, leaving Heatley all but ignored.

The first patient known to get the penicillin therapeutically is recorded in history as occurring on the 12th of February, 1941. The patient's name was Albert Alexander, an Oxford constable, who had a minor scratch that became infected, probably with a mixture of *Staphylococcus* and *Streptococcus*. The sulfonamides had not worked, and the infection had spread, leaving pus oozing out of his suppurating incisions. Penicillin at a dose of 200 mg was provided to Constable Alexander, with supplemental doses of 100 mg at three-hour intervals.

By the 17th of February, Constable Alexander had dramatically improved and was now in stable condition. Unfortunately, the penicillin had run out! The doctors took the unprecedented and seemingly desperate steps to recover the penicillin from the patient's urine, re-administering the recovered medicine for another five doses.

Then, on February 22, 1941, another patient, 15-year old Arthur Jones, with the dread septicemia from a hip surgery, had been provided with all of the remaining penicillin, including that recovered from Alexander's urine. At this point, all of the penicillin was gone. The infection flared up in Jones, but eventually he recovered on his own, without any penicillin for the remainder of his illness. It is clear that the penicillin that had been given to Jones was just enough to turn the tables on the hip infection and permit a full recovery.

Constable Alexander was not so fortunate. With all of the penicillin having been spent and having been treated for only 4 days (it was not enough), Alexander deteriorated, passing away on the 15th of March, 1941.

The first critically ill patient to be conclusively saved by penicillin, a serious case in which death was imminent and which was most decisively reversed by treatment with the penicillin, was the case of Anne Miller. The 31-year old Mrs. Miller had recently suffered a miscarriage and as a complication now had a severe case of the deadly septicemia. It was only through word-of-mouth that penicillin would become available for Miller, as her physician knew someone who knew Florey, who was already in the U.S. attempting to mass produce the anti-bacterial agent. With freshly prepared therapeutic penicillin having arrived at the Yale hospital on the 14th of March, 1942, a Saturday, there were no accompanying instructions on how to use it! With Heatley himself reached by phone after some frantic effort, the proper instructions were provided, and the proper doses were given.

Within 24 hours, Miller's fever had broken, for the first time in over a month. Then, on March 19th, the penicillin had very nearly run out! More frantic calls were made, and more penicillin was shipped. Then, again on March 25th, another round of frantic phone calling and shipping produced the necessary penicillin in order to confer a full recovery. Mrs. Anne Miller lived to be 90 years old.

Chapter 20

OTTO WARBURG

1) If there is one word that strikes fear into the hearts of many, it is cancer. Many scientists have proposed theories regarding the causal factors- what are some of the main theories?

I believe you are quite accurate in your assessment of our fearfulness about cancer. Approximately one in every five human individuals will succumb to cancer and die because of this dreadful disease. Thus, the study of cancer biology and of the immune system, a primary biological defense system against cancer, represent two of the largest research areas in the biomedical sciences.

As there are many types of human cells, tissues and organs which can become cancerous, cancer is an extremely complicated and varied constellation of diseases. Likewise, the progression of a normal cell to a tumor mass, called transformation, can involve a series of complicated and laborious steps in the process. Tumors of uncontrolled cell growth may be benign or malignant.

Benign tumors do not involve a pathological effect on normal cellular structure and function. A malignant tumor, on the other hand, can migrate to other areas of the body (metastasis) and affect the functioning of the cells, tissue or organ in a detrimental manner. The effects of a malignant tumor, i.e., cancer, may be felt on biochemical, cell biological, physiological, molecular biological, and on anatomical levels. Malignant tumors cause significant morbidity and mortality.

In general, there are two main categories for causes of cancer.

The first main category consists of environment- and lifestyle-based factors in nature. There are several certain so-called environmental risk factors in which individuals may be exposed to causative agents of cancer, known as carcinogens.

The second main category of cancer causation has to do with genetics.

There are three main types of carcinogens: chemical, physical and microbial. All three classes of carcinogens mediate their transformative effects at the level of DNA by generating genetic mutations. Mutagenic agents that make nucleotide changes in DNA (mutations) may not necessarily mean that tumors will be generated, as a fair number of biological steps may be required to transform a cell into a tumorigenic one.

Frequently, 'multiple hits' of these types are necessary to establish tumorigenesis. Nevertheless, it is well established that certain carcinogens play major roles in the initial processes of forming individual cancers.

2) Now, who was Otto Warburg and what was his theory?

Dr. Otto Heinrich Warburg was a chemistry Nobel Laureate in 1931 and professor of chemistry and biochemistry and director of the Kaiser Wilhelm Institute for Biology, in Berlin, Germany. Warburg was born on the 8th of October, in the year 1883 in the town of Freiburg, im Breisgau, in the Baden territory of the German Empire. In 1906, Warburg took his doctorate degree in chemistry under the supervision of his graduate advisor, Prof. Emil Fischer, himself a Nobel Laureate, in 1901, for his work on the metabolism of purine and sugar molecules.

Of all of the scientific research conducted by Warburg (and his investigative work was quite extensive), he is known best for two theories, in particular, both of which are of prime importance, even to this day.

Warburg's first theory had to do with his Nobel Prize winning work on cellular and physiological respiration. In this area, starting in 1910, Warburg studied a respiratory chain component that he had called 'atmungsferment' which he believed to be an enzyme that bound iron tightly. His later work showed that his atmungsferment was a two-component protein, cytochrome a/a_3, which came to be commonly known as cytochrome oxidase. Many modern microbiology laboratory students will conduct the cytochrome oxidase test on bacteria.

Warburg further studied additional components of the respiratory chain, such as nicotinamide adenine dinucleotide (NAD^+) in 1906, his so-called 'yellow enzyme' flavin mononucleotide (FMN) in 1932, and the related flavin adenine dinucleotide (FAD^+) in 1936. This oxidative phosphorylation work is still important today because this respiratory system is primarily responsible for generating biological energy used by all living cells.

Warburg's second theory had to do with his studies of cancer, beginning in 1922. Under normal conditions, non-cancerous cells generate energy aerobically in mitochondria from the oxidative breakdown of glucose to pyruvate using glycolysis and then entering the Krebs cycle pathway of metabolism. Under normal conditions of exercise, cells may be low in oxygen and undergo anaerobic metabolism with glycolysis generating lactate.

In Warburg's studies of cancer cells he found that they undergo glycolysis both aerobically and anaerobically; thus, the glycolytic metabolism of sugar even in the presence of oxygen is referred to as the so-called and very well-known 'Warburg effect.'

3) Does anyone know what led him to study cancer and the causal factors of cancer?

It appears that although Warburg did not formally study cancer as a research topic until 1922, he nevertheless had an interest in the problem of cancer early on in his academic studies while a university student. Warburg had long desired to make a contribution to the scientific field of cancer biology. From a logical scientific perspective, Warburg's entry into the study of cancer arose from his work with oxidative phosphorylation and oxygen consumption in cells. From Warburg's point of view, he hypothesized that cancer cells require lots of oxygen in order to grow their tumor masses.

4) What about the role of mutations?

As you and I considered earlier, mutagens are known to play a significant role in carcinogenesis. There are several standpoints to consider with respect to mutations and cellular transformation to tumors.

From a chemical mutagenic standpoint, certain chemical agents will bind to genetic elements on DNA and biochemically alter their nucleic acid structures. The structurally altered nucleic acids then deviate from their regular base-pairing properties to create altered nucleotide sequences within DNA molecules. The altered nucleotide sequences along DNA constitute mutations.

One of the most notorious and well-studied chemical carcinogens is the mutagenic compound called benzo[*a*]pyrene, found in soot and cigarette smoke and definitively demonstrated to be a causative agent in the onset of lung cancer. The stepwise workings of the entire biochemical pathway for benzo[*a*]pyrene as a DNA mutagen has been fully worked out. Interestingly, our liver enzymes help many of these chemical agents to become mutagenic by conducting biochemistry on them and making these chemical agents then readily able to damage DNA as bona fide mutagens, possibly to then become carcinogenic.

From a physical mutagen standpoint, ultraviolet (UV) light and ionizing radiation are the best-known examples of such DNA mutagens. UV light rays are emitted from our sun, and it is these rays that provide suntans, which are actually our response to skin being burned by the UV light—the produced pigmented melanin compound is meant to protect us from further burns. The UV rays generate a type of DNA mutation called a thymine dimer, in which two neighboring thymine bases (TT) in a DNA chain are caused to abnormally bind to each other in a permanently covalent fashion (T—T) by the UV light. The thymine dimer alters the 3-D structure of the DNA chain, and our attempts at DNA repair often result in aberrant nucleotides being put in place of the two thymine bases, generating DNA mutations.

Ionizing radiation is thought to damage DNA. Obviously, the greater the radiation dosage or exposure time, the more potentially serious the DNA damage. The worst-case scenario is the breakage of too many of our DNA chains; such damage is terribly difficult, if not virtually impossible, to repair. Often the cell that's harboring the damaged DNA is destroyed and replaced by new cells. In the rare case that DNA strand breakage is repaired but with altered sequences and the cell manages to live, and the cell's DNA is altered in key genes that function in the cell cycle, then a tumor may be possible.

From the standpoint of microbial infection, certain infectious bacteria and viruses play important roles in tumor formation. Although the full cellular biological pathways for bacterial causes of tumors have presently not been worked out, there are certain established associations between bacterial infection and cancers. For example, certain pathogenic bacteria can be involved in carcinogenesis, like *Escherichia coli* strain NC101, which has been connected to colorectal (colon) cancer, and the bacterium *Helicobacter pylori*, causative agent of stomach ulcers, has been linked to gastric carcinoma (stomach cancer).

The associations between cancers and certain oncogenic viruses are clear. There are tumorigenic DNA-based viruses like Hepatitis B, and the herpes virus called Epstein-Barr virus. Likewise, there are oncogenic RNA-based viruses, like human T-cell leukemia viruses (HTLV-1 and HTLV-2) and Hepatitis C.

Regarding the second main cause of cancer, namely, genetics, individuals may be genetically pre-disposed to cancer based on their genomic DNA. Such individuals may be immune-compromised in that they do not have the necessary genes required to invoke a specific antibody or immune cell-mediated immune response to a tumor.

In general, three primary classes of DNA genes have been implicated to be involved with cancers. The first gene type is called a proto-oncogene. Normally, these genes, which are involved in cell growth, turn on when needed and, importantly, turn off when not needed. An altered proto-oncogene may be converted by mutation to an oncogene, or cancer-causing gene, which is not turned off when needed, leading to uncontrolled cell growth.

The second gene class is referred to as a tumor-suppressor gene, e.g., p53, which normally functions to prevent tumor formation. Genetic mutation of a tumor suppressor gene, however, can certainly result in tumors not being suppressed and, thus, allowed to grow uncontrollably, forming larger tumors.

The last gene class has to do with genes that encode DNA repair systems. Normally, dedicated DNA repair machinery can fix many DNA mutations. However, mutationally altered DNA repair genes can result in aberrant DNA repair machinery. The result is that DNA mutations may not be fixed properly, possibly leading to a buildup of DNA mutations and starting the pathway to tumorigenesis.

The human tumor suppressor genes BRCA1 and BRCA2 represent a well-known example of defective mutations in a DNA repair system encoding proteins that normally participate in controlling tumor growth and in repairing damaged DNA. The BRCA1 protein has been shown to repair DNA that has been damaged by radiation, and the BRCR2 protein has been implicated in so-called recombination repair of severed DNA strands. Everyone has two allelic versions of the two genes. Depending on the nature and extent of the mutations present in one or both alleles of each of the BRCA1 and BRCA2 genes correlates with inability to repair DNA damage and a corresponding inability to suppress tumor growth. This is an active area of biomedical research.

5) What role does oxygen play in cancer and its formation and continuation?

Warburg's interest in the relationship between cancer and oxygen was whether cancer cells mediated oxygen consumption behavior at rates that were higher than in corresponding normal (non-cancerous) cells.

His efforts in this area led him to establish the well-known Warburg effect in which various cancer cells undergo aerobic glycolysis as both normal and cancer cells consume oxygen at similar levels. The primary distinction between normal cells and those of cancer cells, discovered by Warburg, was that cancer cells did not inhibit the glycolytic process in the presence of sufficient quantities of oxygen.

Follow up work on this oxygen finding with cancer cells led to Warburg's discovery that defective or inefficient respiratory machinery was responsible for inability of cancer cells to suppress glycolysis under aerobic conditions. During his studies on cancer cells Warburg found that when agents known to collapse the respiratory chain, like cyanide-containing compounds, glycolysis nonetheless proceeded to occur with oxygen being present. You see, in 1861, Pasteur had observed that oxygen inhibited fermentation. Thus, years later, in 1926, Warburg coined the term 'Pasteur effect' to explain the situation in non-cancer cells, in which oxygen regulated glycolysis.

The findings of Pasteur and of Warburg led to the discovery that because cancer cells grow in the presence of oxygen, they will often work to generate a blood supply to tumor masses by forming dedicated blood vessels to do so. Thus, limiting the oxygenated blood supply to tumors by preventing their blood vessel formation is an active area of biomedical research.

6) Was he ever really acknowledged for his work? And what is the status of his theory today?

In his lifetime, Dr. Warburg was thoroughly acknowledged for his scientific investigations, earning the Nobel, among the accolades, in 1931. Apparently, many of his peers had even nominated Warburg for the Nobel numerous times, even after he had garnered the first Nobel. He also received an honorary doctorate degree by academic officials of the prestigious Oxford University. Warburg was known to be particularly fond of this acknowledgement.

Both of Warburg's principle theories have been elucidated biologically, from a molecular mechanistic perspective. The main respiratory chain components have all been purified to homogeneity, their molecular structures solved, and the flow of electrons mostly traced through each of the electron transport chain components. Of course, many other microbial respiratory chains of extremophiles and of yet undiscovered microorganisms need to be examined, but the fundamental principles have been established, and it was Warburg who made the progress possible so far.

The biological mechanism involved in the Warburg effect has been extensively studied, and many of the biological players have been delineated. In the Warburg effect, glucose enters cells by an enhanced process using glucose transporters, allowing glucose to be metabolized in excess to form pyruvate and later by the Krebs cycle to form citrate, which in turn is utilized to make acetyl-Coenzyme A and fatty acids for membrane synthesis—cancer cells need membranes to surround them.

Cancer cells also lose their requirement for growth factors to grow—cancer cells grow just as well without them. As such, cancer cells use a metabolic pathway called the phosphoinositide 3-kinase/serine-threonine kinase (Akt) system to make excess proteins needed to permit cancer cells to grow unchecked and without the need of a growth stimulant in order to do so.

Recently, a protein was isolated from eukaryotic mitochondria organelles (the so-called energy powerhouses of the cell), and the protein was, interestingly, discovered to be conserved from an evolutionary perspective. The new protein has been referred to as IF-1, which stands for inhibitory factor-1. The IF-1 protein is the actual molecular cellular switch that permits cells to convert from a normal cell oxidative phosphorylation mode to a cancer cell aerobic glycolysis mode. In essence, the IF-1 protein is the 'on-switch' for the Warburg effect in cancer cells. In many tumors cells, IF-1 is produced in overabundant levels.

Both of these Warburg inspired theories are regularly included in textbooks dealing with biochemistry, cell biology and cancer biology. Both scientific areas established by Warburg remain as key areas of active biomedical research to this day. Warburg's wish to contribute to cancer research arose out of his hope that such studies would shed light on the dreadful disease in order to eradicate it someday. Although cancer eradication is a long way off from coming to fruition, his contributions to the field of cancer biology have certainly identified key molecular targets for anti-cancer therapies.

In addition to chemotherapy, surgery, and radiation treatments, new advances are always in the pipeline. One recent development involves obtaining a patient's own thymus-derived white blood cells, known as T-cells, which play a major role in cancer immunity. Then these T-cells are genetically engineered to specifically recognize the patient's particular cancer cells while tolerating normal cells. Next, the engineered specialized T-cells are placed back into the patient where they selectively seek and destroy the patient's cancer while leaving normal cells alone.

7) It seems that cancer can strike many different part s of the body- (brain, lungs etc.) that there might be a need for a different theory for each body part? I mean why would one person get brain cancer, and another person lung cancer (other than perhaps smoking)?

Indeed, just about any of the hundreds of distinctive cell types residing in higher organisms can certainly be transformed into tumor cells. In fact, the National Cancer Institute currently estimates that there are at least over 100 different cancer types, each type potentially with their uniquely distinctive transformation mode of operation.

Many distinctive factors influence the type of tumor a cancer patient may experience. Such factors include one's natural immune system defense status. Another factor includes genetic inheritance that influences susceptibility to certain cancer types. A third factor is the extent of exposure to certain carcinogens. Regarding microbial causes of cancer, the modes of transmission to patients are numerous and may include common ones like food-borne, air-borne, blood-borne, etc.

I think it is interesting that you have inquired about brain cancers because several cellular mechanisms have become known in this regard. The vast majority of brain cancer cells harbor a genetic mutation in the gene encoding a Krebs cycle enzyme called isocitrate dehydrogenase, which under normal circumstances converts the intermediate metabolite isocitrate to α-ketoglutarate. However, in brain cancer cells, the enzyme is defective and instead produces a cancer-specific metabolite, called 2-hydroxy-glutarate. This particular so-called onco-metabolite then results in the alteration in the gene expression behavior in brain tumor cells, allowing them to grow uncontrollably. Again, these onco-metabolites themselves may yet be good targets for anti-cancer chemotherapy.

8) What have I neglected to ask?

Because Prof. Warburg was such a famous investigator, his scientific research laboratory was a highly sought after locale with which many a student of biochemistry and cell biology eagerly wanted to become an integral part. Indeed, Warburg's laboratory housed many excellent students, postdoctoral fellows, and visiting professors, some of whom would later become Nobel Laureates.

One prominent student of Warburg's was the great Prof. Dr. Hans Krebs (Nobel, 1953) for his work in elucidating the tricarboxylic acid cycle pathway. Krebs even honored his great mentor by writing a biography about him—even though Krebs nostalgically reported that Dr. Warburg had once summarily dismissed Krebs when he was still a young research investigator in Warburg's laboratory.

Other protégés of Warburg include Otto Meyerhof (Nobel, 1922) for his studies in elucidating key steps in glycolysis; Fritz Lipmann (Nobel, 1953) for his elucidation of the pyruvate dehydrogenase complex and linking glycolysis to the Krebs cycle; and Hugo Theorell (Nobel, 1955) for his work with enzyme catalysis.

It is interesting that Warburg managed to survive severe persecution by Nazi Germany during the 1930s and 40s. Warburg refused to leave his laboratory or Germany. Krebs has provided an explanation for all of this in his biography of Warburg. Krebs pointed out that because Warburg's mother was non-Jewish, key Nazi officials, Hermann Göring himself among them, deduced that Warburg was apparently not Jewish enough. However, Warburg was closely watched and had severe restrictions imposed on his whereabouts and activities. He was also barred from teaching at university and allowed to do only research. Krebs also pointed out that perhaps German Nazi officials did not persecute Warburg further as they were possibly hoping for a cancer cure to emerge from Warburg's laboratory.

Chapter 21

CARL AND GERTY CORI

1) Professor Varela, in this interview we are going to talk about the Cori Cycle- but first, what do we know about the early days of this couple growing up in Europe?

Drs. Gerty Theresa and Carl Ferdinand Cori were a wife and husband team of scientific investigators who elucidated the so-called Cori Cycle of metabolism and shared the Nobel Prize in the scientific fields of Physiology or Medicine in 1947.

Gerty Radnitz was born on the 15th of August in the year 1896. Incidentally, other sources report that she was born on August 8. Her parents were Otto and Martha Radnitz who lived in Prague, which was located within the so-called Austro-Hungarian Empire, as it was known at the time. Today, the location is known as the Czech Republic. Gerty Radnitz was home-schooled during her elementary school years and was then admitted to a girls Lyceum in 1912. It was during this time-period that young Gerty was influenced to become a medical doctor by a relative (an uncle) who was also a medical pediatrician. In 1914, Radnitz entered medical school at the Charles-Ferdinand University, Prague, graduated with her M.D. degree from the institution in 1920, and married Carl Cori in the month of August during her graduating year. It was during her years spent in medical training where she and Carl had met.

Carl Cori was born on the 5th of December in the year 1896 to parents Carl Isidor and Maria Lippich Cori in Prague, in the Austro-Hungarian Empire. Young Carl was raised in the town of Trieste. Carl entered medical school at the Charles-Ferdinand University (known today simply as Charles University, Prague) during the year 1914, the same year as Gerty. He, too, graduated with his M.D. degree in 1920.

During post World War I years the young married couple encountered hardships, including starvation and anti-Semitism, even though Carl was Catholic and Gerty, though raised Jewish had converted to Catholicism when she married Carl. Nevertheless, these adversities of early 20th century European life compelled them to move in 1922 to the U.S.

where they worked at the so-called New York State Institute for the Study of Malignant Diseases, in Buffalo, New York, nowadays called the Roswell Park Memorial Institute. The young married couple became U.S. citizens in 1928. In 1931, the Coris moved to Washington University, in St. Louis, MI, where they worked together until her untimely death in 1957, ten years after they had garnered the Nobel.

2) Now, I understand they were initially interested in many different realms of medicine. Let's start with the pituitary gland- why were they interested in this particular part of the human body and its function?

The anterior pituitary, among many other functions, regulates a physiological pathway known as the adrenocortical system. In this system, a hypothalamic hormone molecule called corticotropin releasing factor (CRF), also referred to in modern times as corticotropin releasing hormone (CRH), causes the release of ACTH (adrenocorticotropic hormone) from the anterior pituitary during stress or some other physiological stimulus.

The ACTH then causes the hormone cortisol to be released from the adrenal glands, which then cause the conversion of glycogen into free glucose in the liver to mediate stress physiological and behavioral responses.

The Cori research team and their laboratory associates became interested in this so-called adrenal-pituitary axis in order to examine the role of insulin on glucose metabolism. At first, the Coris tested the hypothesis that insulin enhanced the utilization of glucose by circumventing the regulatory influence of the adrenocortical players within the adrenal-pituitary axis.

They had postulated that insulin had an effect upon certain sugar metabolizing enzymes in order mediate this bypass of the regulatory effect of the adrenocortical molecules. Unfortunately, the idea was largely incorrect as it was found later that insulin instead mediated its effects by enhancing the permeability of glucose within rat muscle cells. Later, insulin was discovered to regulate the expression and subsequent activity of glucose transporters within the cellular membranes. This work by the Coris nevertheless established the importance of using purified cellular and molecular players in order to readily and effectively study important physiological cellular mechanisms.

3) Further, they seemed interested in insulin. What advances did they offer in that realm?

Insulin from the beta cells of the pancreas decreases the plasma glucose concentrations by the insulin-receptor pathway in cells from fat, liver and muscle tissue by mediating glucose uptake into cells, thus decreasing blood glucose concentrations. The Coris' interest in insulin was in primarily how its effects on blood glucose levels were mediated by the hormone during the breakdown of the sugar storage form, called glycogen. There were many studies involving insulin throughout the years by the Coris and their laboratory students, postdoctoral fellows and collaborators. One notable finding was that insulin enhanced glucose uptake into muscle cells, presumably by affecting the rate of transport across the membrane by glucose transporter proteins.

The experimental studies of insulin by the collaborative Cori husband and wife team, however, had an appallingly troublesome start.

In his 1969 biographical and scientific memoirs published in the *Annual Reviews of Biochemistry* on the topic, Carl Cori related how the married couple encountered certain difficulties in their efforts to begin work collaboratively together on insulin.

The trouble started soon after Carl had applied for a new academic position in the mid-1920s. He was told by the departmental chair that an offer would be made to Carl but under the following provisos: 1) he take English lessons to improve upon his speaking accent 2) he terminate all studies on insulin and that 3) he and Gerty not work together in order to conduct research. Carl had replied to the potential employer that he (Carl) was certainly happy to adhere to the speaking lessons, of course, but the other two demands were most certainly untenable. During the interview process in which the Coris had visited the unnamed institution, Gerty had been brazenly pulled aside by the hiring committee representatives and had informed her in no uncertain terms that she was an obstacle to her husband's career and, furthermore, that it was considered un-American for husband and wife to work together. Considering all that they had already been through in Europe, her treatment in this altogether new way in a new country was arguably over the top—Gerty had cried tears over the incident.

Frankly, I find it difficult to believe that seemingly intelligent investigators could somehow understand complex biochemical and physiological concepts but needed external help in understanding foreign accents by requiring Carl to improve his speech, for them. Fortunately, for the good of science and the rest of us, the Coris declined the offer.

4) Epinephrine is a very common drug that many use. How did they get involved in what is today often known as the EPI-pin? And what were some of their insights?

The Coris' scientific interests in the hormone epinephrine involved its relationship to the sugar concentrations in the blood; they had entertained the idea that somehow there was an influence to be had by epinephrine in the release of glucose molecules from glycogen stores and movement of glucose in the blood and in the formation of lactic acid. They had called the breakdown of the glycogen stores glycogenolysis, and it had been brought about by the epinephrine.

During the glycogenolysis process they observed that a sugar called hexose monophosphate was produced in frog and rat muscle tissue preparations after the administration of epinephrine. Epinephrine also regulates the synthesis of glycogen. The Coris had followed up on the reverse of the physiological process in which the absence of epinephrine showed the disappearance of the hexose monophosphate with the concomitant appearance of inorganic phosphate and, importantly, of glycogen.

The Coris' seminal studies on the biochemistry and physiology of epinephrine set the stages for the discovery of new insights involving the energetics of metabolism. Epinephrine increases the availability of the sugar glucose for biological energy by promoting glycogenolysis, i.e., the breakdown of sugar storage form glycogen into smaller molecules like glucose-6-phosphate. The Coris purified a variety of sugars, like the glucose-6-phosphate and discovered a key enzyme, called glycogen phosphorylase that functioned during glycogenolysis. They also studied the kinetic behaviors of the enzymes.

Furthermore, the epinephrine hormone stimulates the metabolic pathway called gluconeogenesis (meaning new sugar made), a biochemical process involving the biosynthesis of new carbohydrates from metabolic intermediates, like, for instance, pyruvate.

Although the Coris' work with epinephrine was primarily geared to the metabolism of the various forms of carbohydrates, their early work did lead to the later development of the hormone for many other metabolic and physiological purposes. The Coris did not work directly on the development of the Epipen itself. The basis of the Epipen is to provide the hormone during cases of anaphylaxis after a case of, let us say, an extreme allergic response to a potent allergen. Such strong immunogenic responses can be deadly if not treated right away. The Epipen is supposed to provide a rapid dose of the epinephrine straightaway. Recently, the Epipen was involved in an embarrassing overpricing scandal by its prime vendor.

5) Now, what is glycogen storage disease and why is it important?

Sadly, at almost the very same time that the Coris received word in 1947 that they had gotten the Nobel for their important work of the Cori Cycle discovery, Gerty learned that she had a debilitating and fatal type of anemia, nowadays called myelosclerosis. It was also about this time, incidentally, that Gerty began in earnest her new work on genetic diseases, such as the so-called glycogen storage diseases.

She was primarily interested in the sugar metabolizing enzymes that may be deficient in certain glycogen storage diseases. In the laboratory, Gerty studied various clinical samples of liver and other tissues from patients with assorted glycogen storage diseases. She then purified glycogen from the clinical samples of human liver tissue and muscle tissues.

In one of these forms of glycogen storage disease, called Von Gierke's disease, Gerty determined that the enzyme defect lay in an enzyme called glucose-6-phosphatase, which catalyzed the production of the sugar glucose out of the overall glycogen macromolecule taken from the clinical patients. Depending upon the extent of the severity of the Von

Gierke's disease, the enzyme was either reduced in concentration in mild disease or completely deficient in its levels in fatal cases.

Another one of these forms of the glycogen storage disease is characterized by a defect in the enzyme that de-branches glycogen during its breakdown. The so-called glycogen debranching enzyme, named amylo-1,6 glucosidase, normally works at chemical bonds located at the branch points of the long glucose chains in glycogen, to release free glucose. Gerty had taken to referring to the enzyme as 'The Debrancher.' A defective debrancher enzyme resulted in a toxic buildup of glycogen in critical tissues, like the liver, cardiac muscle or in skeletal muscle.

In 1949, the Coris and colleagues were able to isolate the enzyme called phosphoglucomutase, which was responsible for producing glucose-6-phosphate from the substrate glucose-1-phosphate. The enzyme simply moved the phosphate from carbon number one to carbon number six of the glucose molecule. The biochemical reaction is always included in modern biochemistry textbooks.

Gerty and her laboratory staff worked out how the glycogen breakdown works first with the glycogen phosphorylase, releasing glucose-1-phosphate molecules and then with the debranching enzyme to free up the glucose molecules that were tied up within the glycogen.

The glycogen storage disease type III, which was due to a defective or deficient debranching enzyme in patients, was renamed in Gerty's honor as Cori's disease.

They also purified the relevant glucose-1-phosphate, which later became known as the so-called 'Cori ester.'

In studying another form of the glycogen storage disease, called McArdle disease, Gerty discovered the insight that in the patients' muscle tissue the defective enzyme, glycogen phosphorylase, must be present in two forms as encoded by two distinctive genes within the human genome. She had been correct in her prediction of the presence of the two phosphorylase encoding genes.

As Gerty's condition continued to worsen, she was unable to maintain her enthusiastic and exuberant energy that she had possessed for most of her life and was unable to walk without becoming too exhausted. Carl actually used to carry her from location to location. Soon Gerty had spent less time in the laboratory and stayed home more often, at which point Carl would stay home with her to care for her. Ten years after the Nobel, Gerty passed away on the 26 of October, in 1957.

6) And the Cori Cycle- what insights did they have about this mechanism?

The Coris were instrumental in discovering the Cori Cycle and in elucidating its mechanism of operation. In the cycle, the molecules glucose and lactate (known as lactic acid in some circles) cycle between skeletal muscle and the liver. The Cori research team used adrenaline (i.e., epinephrine) to analyze the metabolism of glucose and glycogen to elucidate the mechanism of the Cori cycle. They studied what happens with these glucose and lactate molecules during extremely intensive muscle activity.

Individuals having undergone intense muscle contractions will continue to breathe heavily, even after the period of intense activity is over, thus using up a lot of the available oxygen in order to undergo a respiratory process called oxidative phosphorylation. The oxidative phosphorylation system in living organisms is a process that shuttles electrons from foodstuffs to the respiratory chain to oxygen, which is the final electron acceptor; meanwhile, an energy charge, called a proton gradient, builds up, as the electrons are transferred along the components of the respiratory chain. This so-called proton-motive-force is itself used as an energy source to make ATP, the main biological energy used by living organisms in order to live.

Next, the ATP formed by this intense biological activity is now used to carry out the process called gluconeogenesis, which makes more of the sugar glucose from non-sugar metabolic precursors, like oxaloacetate (an intermediate in the Krebs cycle) or pyruvate, the end-product of glycolysis.

Finding out what happens next to the newly generated glucose molecules in this process is the insightful knowledge conferred by the Coris—the Cori Cycle.

7) I understand they shared the Nobel Prize- what were they awarded the Nobel Prize with and for what imminent discoveries?

The Coris formulated the Cori Cycle in 1929, and it is this insight that earned them each a share of the Nobel Prize in 1947. The Cori Cycle they discovered speaks to the physiological and biochemical cooperation that occurs between the blood, liver and skeletal muscle.

In the Cori Cycle, glycogen is used to undergo anaerobic glycolysis in order to generate the ATP energy that's needed during intense activity produced by hard-working skeletal muscles, producing lactate as a result. Lactate molecules produced by the glycolysis in the skeletal muscle is then transported through the blood circulation to the liver, where the lactate is then used to make glucose using the gluconeogenesis metabolic pathway. In this gluconeogenic pathway, ATP in the liver is utilized to generate the glucose molecules. The newly made glucose leaves the liver and goes back via the blood to the muscle to resupply it in order to replenish the depleted glycogen stores during the recovery from the exertion.

In summary, during vigorous exercise activity, the Cori Cycle kicks in and involves lactate production from glycogen stores in the muscle; the lactate then goes to the blood to reach the liver, where it produces the glucose. Next, liver glucose goes to the blood to move the sugar back to the muscle; then the cycle starts again by making muscle lactate, repeating the process.

The importance of the Cori Cycle lies in its purpose of supplying glucose sugar as a source of energy for muscle during exercise.

8) What have I neglected to ask?

Many hardships were encountered and endured by Gerty Cori throughout her life. I mentioned above the anti-Semitism she encountered in Europe and the inappropriate treatment she received by Carl's interviewers during his application for a new post. These are, however, only the tip of the iceberg.

First, while a postdoc in Europe in 1920 after the Great War, food shortages were common, and Gerty had developed a condition known as xerophthalamia, characterized by a vitamin A deficiency and blindness. Fortunately, her condition had improved with a proper diet.

During their efforts to move to the U.S. in 1922, only Carl was able to make the migration, while Gerty had to stay behind in Europe, as there was no available research or academic post for her in the U.S. In those days, it was considered inappropriate for a married couple to be separated from each other for an extended period of time.

After six months of separation through no fault of their own, Gerty was finally able to join her husband already in Buffalo at New York State Institute for the Study of Malignant Diseases. Gerty's new post in the department of Pathology grew precarious, however, when she had failed to find any evidence of an amoebic cause of cancer, her supervisor's favorite but untenable theory. As a result, Gerty was threatened with a prompt dismissal unless she ceased working with Carl.

The same institute in New York, and probably many other institutions at the time, categorically frowned upon married couples collaborating together on work, and the Coris became a target of such a policy. A rare exception, however, was made for the Coris, and they began their extremely successful collaborative studies on the Nobel Prize winning Cori Cycle.

In addition to their work on sugar metabolism in cancer, Gerty had also conducted studies examining the effects of X-rays upon tissue, leading historians of science to speculate whether her exposure to X-rays contributed to her fatal anemia.

Prior to the Nobel, Gerty's status was frequently on a sub-par basis compared to that of Carl's. For instance, at the time it was often standard to include the names of expert research personnel on letterhead stationary—Gerty's name was omitted, despite their equal status within the Cori laboratory. In correspondence to the Coris, Carl was consistently addressed as 'Dr. Cori' while Gerty was frequently 'Mrs. Cori.' Her academic credentials were routinely ignored.

In another incident, a letter from an editor of the *Journal of Experimental Medicine* informing Gerty and her co-authors of the acceptance of the manuscript for publication, all authors were addressed as 'Dr.' except Gerty, who was addressed as 'Mrs. Cori,' even though she held a *bona fide* M.D. degree.

Even though Carl and Gerty Cori were a team and worked together as equals in their research laboratory, his career rapidly moved through the academic ranks and with much greater speed than did hers. In fact, Gerty's full academic rank as professor did not transpire until the Nobel, whereas Carl had been made full professor and departmental chair many years prior to the Nobel.

After receipt of the Nobel, the Cori research laboratory became a magnet for many excellent and eager young students wishing to study science under the Coris supervision. Many colleagues who were fortunate to have studied science in the Cori laboratory went on to have quite prominent careers of their own. For example, Arthur Kornberg, who earned a Nobel in 1959 for his discovery of DNA polymerase, attributes his success to the Coris' glycogen phosphorylase enzyme, and not necessarily DNA base pairing, which led him to his discovery of the Nobel-winning DNA polymerase.

Others who were fortuitous enough to enjoy the training and inspiration of the Cori lab include the following Nobel Laureates: Christian de Duve, discoverer of sub-cellular fractionation technology; Severo Ochoa, discoverer of RNA polymerase, a key enzyme for transcription; Luis Leloir discoverer the role of sugar nucleotides in the synthesis of carbohydrate molecules; Edwin Krebs (unrelated to Hans Krebs), discoverer of the enzyme phosphorylase kinase, a key enzyme in the conversion of phosphorylase *b* to phosphorylase *a*; and Earl Sutherland, discoverer of cAMP (cyclic adenosine monophosphate) as a regulator of sugar metabolism.

Many colleagues have written rather fondly about their experiences in the Cori laboratory throughout the years. Students, postdocs, collaborators, visiting scientists, etc. have all conveyed essentially the same motif within the Cori laboratory of hard work, dedication, and especially of the excitement to be had from learning new facts on thresholds of the biomedical sciences, like biochemistry and physiology. As a whole, their scientific work was productive and far-reaching. Indeed, the lifework performed by the Coris and their colleagues are still completely relevant today.

It is written that, after Gerty's early death at 61 years of age, Carl did not take it well. After she died, he was not surprisingly devastated by the loss and took an extended leave of absence from the laboratory. He eventually remarried, and his new marriage to Anne Fitzgerald Jones was known to be a happy one. After his retirement in 1966, Carl became a visiting professor at Harvard Medical School, in Boston, and continued working in a lab housed at the famous Massachusetts General Hospital until his death on the 20th of October, in 1984, at 87 years of age.

Chapter 22

ERNST CHAIN, HOWARD FLOREY, AND NORMAN HEATLEY

1) Professor Varela, today we are going to discuss three great scientists, who, apparently working together, "purified" penicillin. First of all, what do we know about the backgrounds of these three individuals?

You have chosen an excellent topic for this chapter in the series. The fact that the important antibiotic penicillin was discovered by the amazing scientist Sir Alexander Fleming is virtually universally known. Not many people, however, are aware of the amazing biochemists and of their tremendous efforts they took in purifying the penicillin. The individuals who were key to the scientific achievement in which they successfully accomplished the production and isolation of penicillin were Drs. Howard Florey, Ernst Chain, and Norman Heatley. Their efforts represent a critical advancement in the effort towards an efficacious treatment against several serious bacterial infections. Their work resulted in a significant historical first in chemotherapeutic treatment of infectious disease.

Let's consider first the background of their principal leader of the penicillin isolation project, Dr. Howard Walter Florey. He was born in Malvern, near the suburb of Adelaide and part of the city of Unley, in southern Australia, to parents Joseph and Bertha Mary Wadham Florey, on the 24th of September, just prior to the entry into the 20th century, in 1898. Joseph was a businessman who dealt in the production of boots. Bertha was the second wife of Joseph after his first wife, Charlotte Ames, died of tuberculosis, frequently called the Great Consumption during the time. Together, Charlotte and Joseph had two daughters, Charlotte and Anne. Bertha and Joseph had three offspring, two older sisters, Hilda and Viletta, plus the youngest, Howard.

As a child, Howard Florey attended Kyre Preparatory College, a private school, in Adelaide, where he was given the nickname "Flos," and in 1911 he moved to St. Peter's Collegiate School, also in Adelaide. After graduation in 1916, Florey enrolled in medical school, housed at Adelaide University. As a medical student, Florey's father Joseph died in September of 1918, leaving the family in dire financial straits. Nonetheless, Florey managed to graduate with his B.A. degree in surgery by December of 1921, earning both excellent grades and a Rhodes scholarship to Oxford, England, where he had moved to in January of 1922.

While at Oxford, Florey studied physiology, which is the study of the function of living systems. At Oxford, Florey studied under the noted neuroscientist Charles Sherrington and earned an M.A. in physiology, in 1924. Florey received another scholarship to attend Cambridge, England, where in 1924 he entered the laboratory of immunologist and professor of pathology, Henry Dean. Flourishing in Dean's laboratory at Cambridge where he studied the pathology of blood circulation in the brain and mucous secretion from the colon, Florey published four papers in prominent scientific journals. In 1926, Florey became a research fellow at London Hospital, and in October of that year, he got married to Mary Ethel Hayter Reed. In 1927, while taking a lectureship post in pathology, Florey earned his Ph.D. degree from Cambridge, where he studied the circulatory flow of lymph and blood.

Dr. Florey spent brief intervals at Oxford and Cambridge before moving to Sheffield to take a professorship in pathology in 1932, and later, in May of 1935, Dr. Florey finally became a professor of pathology at Oxford, where he was to do his penicillin work.

The second member of the penicillin production trio, Ernst Boris Chain, was born on the 19th of June in 1906 in Berlin, Germany. Chain's mother, Margarete Eisner, was German. His father was Michael Chain, a Russian industrial chemist who worked in a manufacturing plant for chemicals, in Berlin. Chain attended a gymnasium called Luisen in Berlin. The child Chain often visited his father's laboratory, becoming interested in chemistry, and in 1919 when young Chain was only 13 years old his father passed away, leaving the immediate Chain family in terribly difficult financial circumstances—a situation similar to that seen with Florey. However, Chain was also able to go to university as an undergraduate, at the Friedrich Wilhelm University, Berlin, Germany, where he learned five languages and was a talented pianist, even seriously considering becoming a concert pianist for a career. Chain took his undergraduate degree in chemistry and physiology from Wilhelm in 1930.

With his undergraduate degree in hand, Chain worked briefly in the physical chemistry department at the Kaiser Wilhelm Institute and at the Charité Hospital, in Berlin, where he is reported to have earned his first Ph.D. degree, in 1930. Being from a Jewish family, Chain left Germany in April of 1933, during the rise of Hitler's Nazi Germany. Arriving practically financially destitute in London, England, Chain was barely able to secure tenable work at University College Hospital Medical School under the direction of the J.B.S. Haldane, a distinguished physiological geneticist. Working briefly in the laboratory of Charles Harington in May of 1933, Chain and Harington did not get along well.

Shortly afterwards, in October of 1933, Chain was accepted into graduate school in the Ph.D. program at Cambridge University, working in the biochemistry research laboratory of Prof. Frederick Gowland Hopkins, housed at the Sir William Dunn School of Biochemistry, where Chain took up the study of phospholipids and enzymes. Hopkins was under the impression that earning another advanced degree, at Cambridge, would help keep Chain out of having to return to Germany, where it was considered dangerous.

Chain and Hopkins got along quite well. Chain started his studies on snake venom and its effects on the nervous system, focusing on the biochemistry of the venomous enzyme involved in causing paralysis. Completing his second graduate thesis, Chain took his Ph.D. from Cambridge in 1935.

Hired by Florey, who needed a biochemist, in September of 1935, Chain became demonstrator and lecturer of chemical pathology at Oxford University in the Sir William Dunn School of Pathology, and in his new laboratory he completed his studies on the biochemistry of snake venom. It was towards the end of 1935 when Chain learned that his mother and sister, Hedwig, still in Germany, had been sent to Theresienstadt, a concentration camp in Czechoslovakia. In 1936, Chain suffered a nervous breakdown and during his recovery in Paris he was introduced to Anne Beloff, a prominent biochemist in her own right and who was later to become Chain's wife in 1948. In 1939, Chain became a citizen of England. In 1942, he had learned that his mother and sister perished in the Nazi holocaust at Theresienstadt.

The third penicillin purifier, Norman George Heatley, frequently referred to as "penicillin's forgotten man," was born as an only child to parents Thomas and Grace Heatley on the 10th of January in the year 1911 in Woodbridge, Suffolk, in England. Heatley's father was a veterinary doctor specializing in farm animals. Heatley was reported to have accompanied his father on his rounds, and he (Norman) was impressed with his father's skill with animals; importantly, however, the young Heatley was deeply impressed with his father's hobby of micro-manipulation skills in fixing broken china tea cups and saucers. This experience was to become a long lasting interest for Heatley. The child Heatley attended St. Felix grammar school in the adjacent town of Ipswich, England and then moved to Folkstone, along the coast of England to attend a boarding school called Westbourne House.

In 1929, Heatley graduated from high school and enrolled in St. John's College, in Cambridge, England. Concentrating in natural sciences, Heatley took his undergraduate degree in 1933. Heatley then studied biochemistry for an academic year after his college graduation, at which time he became a laboratory assistant to Prof. F.G. Hopkins at Cambridge University. Next, Heatley entered graduate school at Cambridge and studied micro-chemical techniques in the laboratory of Joseph Needham and graduated with his Ph.D. degree in 1936. It is during this time that Heatley's expertise was noticed by Chain, and upon Chain's recommendation, Heatley was hired by Florey at Oxford.

2) What brought them together? And where did they work together?

The trio of penicillin makers, Florey, Chain and Heatley, were housed first at the Sir William Dunn School of Pathology, in Oxford, England. Florey had become professor of pathology in May of 1935, and at Oxford he first hired Chain, who arrived in September of that same year. The following year, in 1936, Florey hired Heatley, who arrived at Oxford almost exactly a year after Chain.

All three individuals were very highly recommended by their immediate superiors, and all of them brought to the table badly needed expertise. Florey supplied facilities, funding, effective leadership, and stability amongst the other two seemingly strong-willed and completely divergent colleagues. Chain provided expertise and knowledge of biochemistry, metabolism, fermentation, and cell culturing. Heatley, also a biochemist, delivered expertise in the assembly of apparatuses, scaling-up of techniques, bacteriology, testing of penicillin's efficacy, and especially of micro-manipulation of equipment. In particular Heatley was key to solving a problem with the instability of the penicillin during purification.

Later, Florey, Heatley and other personnel were to move to the U.S. to complete their large-scale production of penicillin, in relative safety from the German blitz bombing. Here, legend has it that Florey, Heatley and others actually smeared the penicillin-producing mold spores on the insides their sewn coats as they travelled to the U.S., in order to remain incognito and, importantly, to keep penicillin out of German hands, should they be captured.

There is a famous quotation by Prof. Henry Harris, Florey's successor at Oxford in later years. Harris is reported to have summarized the contributions of the four main penicillin players as such, "Without Fleming, no Chain or Florey; without Chain, no Florey; without Florey, no Heatley; without Heatley, no penicillin."

3) Now, can you walk us through this process of "purifying penicillin" and review the initial stages of discovery and then the need for this purification process?

The first attempts to purify penicillin from the *Penicillium notatum* mold were performed by Dr. Alexander Fleming, himself, shortly after his famous discovery, in 1928. Several other groups also tried to isolate the penicillin, and all failed to adequately produce enough for practical purposes. There were essentially three main focuses: growing the mold, extracting and purifying the penicillin from the cultured mold, and the testing of the purified penicillin on bacteria.

Let's consider the mold growth efforts, first. On Petri plates, the *P. notatum* mold formed large solid mats, and on top of these mats secreted penicillin emerged from the mats forming little droplets of liquid juice harboring the penicillin. The penicillin juice could be taken up with syringes and tested on bacteria. The problem here was that this method was not practical for large-scale production and the secreted penicillin juice was found to contain over 50 other non-necessary components. Obviously, it was important to have penicillin in a pure state. Efforts turned to large-scale culturing of the *P. notatum* mold, using a variety of broth medium nutrients, pH, temperatures, time-periods, etc.

As the mold culturing conditions were being optimized, the problems of small- and large-scale penicillin extraction and isolation were addressed, as well. The first step involved filtration of the grown mold, addition of acid and cooling of the filtered acidified mold juice. Next, the chemical ether (or sometimes amyl acetate) was added to the acid mold juice—the penicillin would move into the ether. Next, using Heatley's instability problem-solving idea, a mixture of alkaline and water was added to the ether-penicillin mixture, shaken vigorously, allowed to separate and distilled. The penicillin moved to the basic water and in stable form. Next, the penicillin solution was freeze-dried into a powder.

Next, the technique was moved to a large-scale process, using an automated apparatus to filter, add acid, alkali, and other reagents, etc. Interestingly, the large-scale production apparatus was assembled by the trio using make-shift and rudimentary materials, like bathtubs, trash cans, and even milk churns, in order to produce the penicillin.

Lastly, the anti-bacterial nature of the purified penicillin needed to be tested against bacteria, to see if it killed any of them. The first experiments of purified penicillin involved making Petri plates with agar growth medium in them. The bacteria targets were inoculated onto the Petri agar plates. Next, short hollow cylinders, made out of glass, and resembling extremely short straws, were placed on top of the inoculated agar, and the purified penicillin was poured into the cylinders. The pure penicillin would then emerge from the cylinder vessels and diffuse into the agar where the bacteria were located. If the purified penicillin killed or inhibited the growth of these bacteria, then a zone of growth inhibition would form about the cylinders in the agar medium. No bacteria were found in these zones where the penicillin moved into around the cylinders. The presence of these bacteria-free zones were good indicators of the effectiveness of the pure penicillin in killing bacteria.

4) Today, in the year 2018 - what are the challenges in terms of constructing if you will, this miracle drug we call penicillin?

After the large-scale production of penicillin by Florey, Chain and Heatley, in the mid-1940s, the miracle occurred of actually surviving serious bacterial infections, for the first time in history. Other antibiotics were discovered, purified and used clinically, such as the tetracyclines, streptomycin, etc., and they were effective against bacterial infection, at first.

A major challenge then (as well as today) was finding new ways to chemically modify the penicillin platform so that the new derivatives retained their stability and anti-bacterial effectiveness. Current challenges continue to fall into the category of ensuring that the new derivatives are safe, and this involves extensive cellular and bacteriological testing and clinical trials, all of which constitute a financial burden.

Sadly, the miracles of penicillin and other antimicrobials, have been compromised. A primary challenge has been that the bacteria evolved to resist the antibacterial activities of the penicillin group of antibiotics and other antimicrobial agents. In fact, new bacterial variants have emerged in which they are resistant to multiple antibiotics. A good example is the penicillin derivative called methicillin, which initially provided an effective treatment of the dread *Staphylococcus aureus* bacteria, a causer of a wide variety of infections in humans and other animals. Soon after the clinical implementation of methicillin, variants of the *S. aureus* emerged from patients who failed to respond to the new medicine. Shortly after the discovery of the methicillin-resistant *S. aureus* (MRSA) clinical isolates, it was further found that these MRSA bacteria were also resistant to multiple antimicrobial agents. Today, MRSA constitutes a serious public health concern.

Present in virtually all penicillins and derivatives, is the active chemical site, called a β-lactam ring structure, which is necessary for the killing effect of these antibiotics, which are oftentimes called β-lactam antimicrobials. This is because penicillin, its derivatives, and other antimicrobial agents that contain the active β-lactam ring structure. Another challenge is the discovery of bacteria, some from soil, that have super enzymes that biochemically destroy penicillin and its derivatives, like β-lactams, with super high destructive efficiencies. These bacteria have specialized enzymes called extended-spectrum β-lactamases (ESBLs) that may confer upon these bacteria the role of the superbug status.

One challenge has been allergic reactions to penicillin and other β-lactams. Penicillin allergies can be potentially life-threatening if an anaphylaxis reaction occurs. One well-known mechanism of allergy to penicillin involves first the drug binding to a person's proteins forming conjugates (penicillin bound to host self-proteins), which induce helper T-cells to activate B-cells, causing them to make lots of a specific type of antibodies (their technical term is class E immunoglobulin, or IgE) against penicillin. These antibody-penicillin complexes stimulate a person's mast cells to produce a set of mediators that cause a person to experience anaphylaxis, which may be mild, characterized by hives, or severe, as in the case of anaphylactic shock, characterized by a drop in blood pressure, shock, difficulty in breathing due to airway constriction, and a swelling of the epiglottis, causing suffocation. Thus, anaphylactic shock can be lethal.

Furthermore, for every class of antimicrobial agent in use today, there are bacteria that are resistant to each of these antibacterial agents. Interestingly, and perhaps disturbingly, use of an antibacterial agent not only selects for bacteria that harbor resistance to that agent used, it actually selects for bacteria that are multiple drug resistant. What's more, genes encoding these multidrug resistance elements may be transferred to other species of bacteria, like in the human gut, in soil, or in sewage, etc., making resistance spread around.

Both investigators and clinicians are actively fighting back against multiple antimicrobial resistances from bacteria, by searching for new antibiotics with novel ways of killing bacteria and in invoking prudent and proper stewardship practices in the uses of anti-bacterial agents. These efforts are of primary importance in hospitals, the environment, and in agriculture. Efforts are aimed at modulating these antibacterial resistance mechanisms, in order to restore their antibacterial effectiveness. Therefore, study of the biochemistry and molecular biology of the antimicrobial resistance mechanisms will continue to be important areas, in order to find new ways to circumvent bacterial resistance. There is so much work that still needs to be done in each of these areas.

5) Were these three men recognized during their lifetimes? What awards did they win, if any?

Of all four players in this story, Alexander Fleming was to receive the vast majority of the accolades. In 1945 the Nobel Prize in Physiology or Medicine for this amazing and important history-changing discovery went to Fleming, for his discovery of the penicillin, and to Florey and Chain, for their production of the penicillin and permitting its clinical use. Others were left out of the Nobel, namely, Heatley, who was instrumental in so many ways for penicillin production, and many others who helped to determine penicillin's structure, its mode of action, biosynthetic pathways, mechanisms of resistance, etc.

Florey, Chain and Heatley were relatively recognized in various ways, with numerous awards, knighthood (to Fleming, Florey and Chain but not to Heatley), and honorary degrees. There were, apparently, varying degrees of disappointment in the recognition received amongst the trio of penicillin producers. First, Florey and Chain, despite becoming Nobel Laureates, continually felt that they did not get the full credit for their efforts that they felt they had deserved. Heatley received the least recognition and credit of all three. It was to profoundly affect Heatley for many years to come.

6) What have I neglected to ask?

I thought it would be informative to our readers to provide an overview of the various cellular molecular strategies that bacteria have evolved to resist the killing effects of penicillin and β-lactam antibacterial agents. These mechanisms have been problematic in the efforts exerted to treat infections caused by bacteria which harbor these resistance factors. The presence or absence of these resistance factors may quite literally determine the death or life matters in patients with bacterial infections.

One of the first β-lactam resistance mechanisms to be discovered was the enzymatic inactivation mechanism in which an enzyme, typically called β-lactamase, cuts the antibiotic into a form that doesn't kill bacteria. Thus, the bacteria live in the presence of (i.e., resist) the antibiotic. Some of these enzymes can destroy large quantities of the antibiotics.

Another resistance mechanism involves the alteration of the target of the antibiotic. These targets are known as either penicillin-binding proteins (PBPs) or transpeptidases, which are important components of the bacterial cell wall building machinery. Penicillin will bind to these targets and prevent their function. As a consequence, the bacteria either fail to make cell walls or produce weak cell walls, making the bacteria die by explosion. The altered target mechanism reduces the penicillin binding, and the bacteria live with their cell walls left intact.

Along these lines, the penicillin target may be made in a great abundance, resulting in lots of target molecules left alone by the penicillin. With the drug target left untouched, the bacteria synthesize their cell walls, letting the bacteria live.

Another effective resistance mechanism prevents entry of penicillin and other antibiotics into the insides of the cell; this bacteriological mechanism is called simply reduced permeability or entry, and prevention of access to the drug target. If the antibiotic cannot get to the cytoplasmic insides of a bacterium, where the targets are, then the bacterial transpeptidase targets successfully make their cell walls, and the bacteria live.

One last resistance mechanism includes an export apparatus called an antimicrobial efflux pump, which is embedded in the bacterial membrane and which sends any internal antibiotic to the outside of the cell, where there are no targets. These antimicrobial efflux pumps may send out only one type of antibiotic or send out several different antibiotic types. These so-called multidrug efflux pumps are studied in my own research laboratory.

Chapter 23

FRITZ LIPMANN

1) Professor Varela, Fritz Albert Lipmann is another famous biochemist- and he apparently shared the Nobel Prize with Hans Krebs. What do we know about Lipmann and his early life and education?

On the 12th of June, just prior to the start of the 20th century, in the year 1899, Fritz Lipmann was born in Königsberg, which was at the time the capital city of the small country of East Prussia, later becoming part of Germany and which is now located in present day Kaliningrad, in Russia. Lipmann's mother was Gertrud Lachmanski Lipmann, and his father, Leopold Lipmann, was a lawyer. Lipmann attended elementary and high schools (called gymnasiums at the time) in Königsberg, where he studied Greek and Latin.

Main influences in Lipmann's early childhood include his older sibling by two years, Heinz; an uncle, a pediatrician who died young due to a burst appendix; and his father, who confessed to young Fritz that he (Leopold) was not a good enough crook to make an excellent attorney. Speaking of confessions, Lipmann reveals in a memoir that he was not a very good student in gymnasium or in university.

After graduation from gymnasium, Lipmann began the study of medicine in 1917, but shortly thereafter his studies were interrupted when he was called into the Army where he conducted his medical service. At that time, the Great War (World War I) had just begun. During this time, he gained experiences in military medicine during battles and observed the Great Influenza pandemic, killing an estimated 30 million people.

With the Great War over in 1919, Lipmann and Heinz then relocated to the University of Munich, Germany, where Fritz had signed up for one academic semester, studying medicine. Lipmann then moved to the University of Berlin to finish his medical degree, in 1924. Upon the advice of his parents, Lipmann spent several months working with cadavers and microscopes, in order to better prepare him for medical studies.

However, doubts about practicing medicine began to creep in; he felt somewhat uncomfortable charging people money for keeping them healthy in return. Lastly, Lipmann took a course in Biochemistry, the study of the chemical reactions in living systems—his lecturer (professor) was Dr. Peter Rona. These experiences led Lipmann to pursue the study of biochemistry.

In his last year of medical studies, Lipmann entered Rona's research laboratory, housed at the University of Berlin. Prof. Rona was an eminent investigator; he had collaborated with another famous biochemist, Leonor Michaelis (enzyme kinetics), and Rona had taught future Nobel Laureates Ernst Chain (penicillin production), and Hans Krebs (intermediary metabolism). Anyhow, it is in Rona's laboratory that Lipmann published his first paper in a scientific journal, in 1924, presenting his results on colloidal chemistry. This work was also to be used for a thesis project, which was required for his medical education. He graduated in Berlin with his M.D. degree in 1924.

In Rona's laboratory, Lipmann also published his results from his experiments working with rabbits in which he had injected with various carbohydrates, such as glucose, starch and glycogen and measured their blood glucose levels, in 1924 and 1925.

With his M.D. degree in hand, Lipmann was still greatly interested in biochemistry and research. Thus, in order to continue with this new interest in biochemistry, Lipmann felt he needed to learn basic chemistry first. Therefore, he had taken a post-graduate fellowship at the University of Amsterdam, in the pharmacology department, where he learned about the main concepts pertaining to biochemistry, in the laboratory of Dr. Ernst Laqueur. Next, he moved back home with his parents and studied chemistry for three years under the supervision of Dr. Han Meerwein, who was a prominent organic chemist at the University of Königsberg.

Dr. Lipmann then made the unprecedented move of entering graduate school for yet another advanced degree, this time the Ph.D. degree. In 1927, he was accepted with no pay at the Kaiser-Wilhelm Institute for Biology in Berlin, directed by the great scientist and biochemist of the day, Prof. Otto Warburg. At this Institute in Berlin, graduate student Dr. Lipmann entered the research laboratory of yet another famous biochemist of the time, namely, the lab of the great Prof. Otto Meyerhof, who had been a protégé of Prof. Warburg. Meyerhof had already received the Nobel Prize in 1922. Warburg was to receive the Nobel later, in 1931.

In the Meyerhof laboratory in 1927, Dr. Lipmann first conducted experiments dealing with measuring the effects upon rabbit blood sugar of yeast extract (autolysate) that was hypothesized to harbor an elusive "glucose activator," which they had hoped would be an enzyme, called hexokinase, contained within the autolysate. Unfortunately, there was no effect on blood sugar levels.

The second project conducted by graduate student Lipmann in the Meyerhof lab involved determining whether the muscle contraction-inducing substance called acetyl choline caused the breakdown of creatine phosphate (denoted as C~P and which is used up to provide energy during acute vigorous exercise) in frog muscle. Unfortunately, acetyl choline had no effect on C~P hydrolysis, although Lipmann did manage to show that C~P was indeed split in half to form individual creatine and phosphate during the contraction of frog muscle. Incidentally, it was Lipmann himself who was the very first to invent the "~" symbol, called the "squiggle" to denote as high energy phosphate bonds. Breaking these energy-rich bonds releases the energy stored in them.

The third and last graduate project performed by Lipmann in Meyerhof's lab entailed the examination of fluoride upon sugar breakdown, a process known as glycolysis. Lipmann demonstrated that fluoride inhibited glycolysis, an important finding. He also showed that the fluoride combined with a form of hemoglobin called met-hemoglobin to form fluoro-met-hemoglobin.

Each of the three projects were published as three separate papers in scientific journals and served as a basis for Lipmann's graduate doctoral thesis, which he earned in 1929. At this time, Lipmann now had both an M.D. and a Ph.D., both of which firmly established his entry into the important field of biochemistry and its sub-field, metabolism.

Lipmann then entered the laboratory of Albert Fischer at the same institution in 1930 studying the growth of cells and using manometry to measure oxygen uptake into the grown cells. But the early 1930s were precarious years as Hitler's rise to power became established. Fortunately, Lipmann and his new wife Freda Hall, an American, were able to leave Germany to accompany Fischer as he moved to his new post in Copenhagen, Denmark.

While Fischer's new laboratory was still being constructed Lipmann travelled to the U.S. for the first time and worked briefly with Phoebus Levene at the Rockefeller Institute for Medical Research, in NY. Working with Fischer, Lipmann studied the "Pasteur Effect," a phenomenon in which facultatively anaerobic cells turn off ethanol fermentation when the cells are grown in the presence of oxygen. It is in Copenhagen where Lipmann converted from the study of animals to the study of bacteria.

Lipmann spent the summer of 1931 at the famous Marine Biological Laboratory (MBL) in Woods Hole, MA, where he shared a laboratory with Leonor Michaelis, another famous biochemist who studied enzyme kinetics. Many biochemists and later physiologists and cell biologists would spend each summer at the MBL in Woods Hole conducting experiments and holding seminars daily.

During the mid- and late-1930s, the situation with Hitler's fascism grew worse, and now Denmark was deemed unsafe, as well. Fortuitously, Lipmann's work on the Pasteur Effect got the attention of Dr. Vincent du Vigneaud at New York's Cornell Medical College in the U.S. In 1939, Dr. du Vigneaud offered Lipmann a temporary post at Cornell, and Lipmann eagerly accepted.

In 1941, Lipmann moved Boston, MA to work as a research associate at the prestigious Massachusetts General Hospital, affiliated with Harvard Medical School. It is here at Mass General where Lipmann conducted his famous Nobel Prize worthy work with Acetyl-CoA, a central metabolite. At Harvard Medical School, Lipmann moved through the academic ranks, first as a research fellow and finally as a full-professor in the biological chemistry department.

Lipmann stayed at Harvard Med till 1957, when he moved to Rockefeller and eventually became professor emeritus. His later works concerned studies of the metabolite carbamyl phosphate and its relationship to the biosynthesis of another metabolite called citrulline, plus works on sulfur biochemistry, and later studies pertaining to protein synthesis, or translation. He passed away on the 24th of July in 1986, at the age of 87.

2) Professor Varela, we know that there are all kinds of biochemical reactions going on in the human body- we have oxygen going in, we may have just had lunch or a nap- Can you tell us about some of the MAIN biochemical reactions going on as we sit here and type and speak?

Well, Prof. Shaughnessy, after just having finished a meal consisting of food substances, like let's say, carbohydrates, protein, some fats, and nucleic acids, an interesting process takes place as these substances are broken down in order to make energy. It may be somewhat of a surprise to you that just about everyone's favorite food consists of electrons—at least they could be our favorites, being that all living beings, from bacteria to humans, have devoted a great deal of our biological metabolic machinery to the acquisition of electrons.

First, the electrons that surround the atoms in the meal are taken away from the food breakdown products, a process called oxidation (electron loss) or oxidative degradation, also called simply catabolism. Next, the electrons taken from food are then brought to the oxygens that you had mentioned were going in. Along the way, the electrons that had been extracted from the molecules making up your meal are shuttled along a series of specialized proteins, collectively called a respiratory chain or an electron transport chain, embedded in the inner membrane of a eukaryotic cell's mitochondrion (an energy powerhouse) or in the membrane of bacteria.

As the electrons are shuttled along the respiratory chain, a bunch of charged hydrogen atoms (i.e., H^+) called protons are pushed out through the chain, building up on one side of the membrane and forming a so-called proton gradient. This proton gradient itself is part of a biological energy form, called a proton motive force, that's used to make another biological energy form, called ATP (adenosine triphosphate). Overall, this metabolic system is called oxidative phosphorylation. The oxidative part is the electron removal from food breakdown products and specialized electron carriers called NADH and $FADH_2$; whereas the phosphorylation part is the addition of phosphate to ADP to make the ATP.

The catabolism of foodstuffs involves a series of biochemical pathways, depending on which food has been consumed by the living being. The breakdown of carbohydrates, like sugars called sucrose, fructose, glucose, etc., involves a metabolic pathway, most often called glycolysis ("glyco" means sugar, and "lysis" means breakdown). Sometimes glycolysis is also called the Embden-Meyerhof-Parnas (EMP) pathway, or even the more obscure designation, the fructose bisphosphate aldolase pathway. My favorite is the EMP pathway, because it is named after three key investigators, Gustav Embden, Otto Meyerhof, and Jacob Parnas, who participated in figuring out the centrally important metabolic pathway.

Interestingly, all living beings on Earth, from bacteria to humans, undergo glycolysis in much the same way. It is a process that was invented by our common ancestors and is unquestionably of major importance today for life.

The breakdown of fats occurs by first breaking them in half, forming a 3-carbon glycerol backbone and, depending of the type of fat, a group of fatty acids. The fatty acids are long chains of carbon hooked up to a bunch of hydrogen atoms. The glycerol is taken to glycolysis and finished up there while the fatty acids are taken apart two carbons at a time, in a process called β-oxidation, to form an extremely important metabolic intermediate, called acetyl-Coenzyme A (Ac-CoA), which then goes to the Krebs cycle.

The breakdown of protein involves the clipping off of the amino acids that make up the protein chains. The amino acids are then further taken to glycolysis and Krebs cycle. Likewise, the breakdown of nucleic acids also involves their movement to glycolysis and Krebs cycle.

Overall, the process of breaking down these food materials in the presence of oxygen is called aerobic respiration. In certain living beings, like bacteria for instance, some of which also undergo aerobic metabolism, they sometimes undergo catabolism without oxygen, a process called anaerobic respiration. Living beings that do their metabolism with or without oxygen are called facultative anaerobes. Besides respiration (whether aerobic or anaerobic), certain life forms undergo a process called fermentation, which involves glycolysis and then any one of a large repertoire of fermentative pathways.

The other half of metabolism, catabolism being the first half, is called anabolism. Here in anabolism, the energy store in the ATP molecule is used to build larger molecules, some being called macromolecules, all of which are critically useful in order for life to occur. One central theme in these metabolic systems is that they use enzymes, the substrates that the enzymes do their chemistry upon, and products that result from the biochemical processes carried out by these enzymes. In fact, several of the key biochemical steps in glycolysis are in play and are shared during some anabolic reactions, but are run in reverse directions, even sometimes using the same enzymes.

3) What can go wrong with these reactions? Obviously, if we do not get enough oxygen, the heart and brain start to die (and we start to die also).

Clearly, you have identified a key biological requirement for many living beings, namely, the necessity for oxygen. When living beings metabolize their meals, like I've mentioned above, the electrons that are taken away from the metabolic intermediates are frequently sent to oxygen. In this sense, the final electron acceptor, if you will, is oxygen.

If one does not get enough oxygen in organisms that are obligate or strict aerobes, the cells, tissues, organs, like the heart or the brain, etc. will die, and consequently the organism will die, as well. In these aerobe organisms, other forms of oxygen, such as carbon dioxide (CO_2) or carbon monoxide (CO), are not good enough to serve as the final electron acceptors—in these cases, only oxygen (and in the form O_2) will suffice.

Another way in which these reactions can go wrong, is when there is, e.g., a lack or deficiency of a substrate, or a missing or malfunctioning enzyme that has been produced in a living being. In the first case, substrate deficiency, it may be the result of a poor diet; and in the case of a dysfunctional enzyme, it may be due to a genetic disease in which the gene that encodes the enzyme has been altered by an inheritable mutation. Perhaps another gene encoding the synthesis of a needed substrate for another enzyme is defective, causing a loss of the needed product.

In some cases, missing or non-functional enzymes mean that a substrate builds up to abnormal levels, causing illness. An example is glucose-galactose malabsorption, in which key transporter proteins that allow sugar entry into cells are lacking, or in the case of lactose intolerance where key enzymes in the metabolic pathway that catabolize lactose sugar is present in low levels. These metabolic conditions may require careful diet or replacement therapies.

4) Now, glycolysis - what is it and how does it fit into chemical reactions?

Glycolysis involves the oxidative breakdown of sugar, like glucose molecules, to produce the metabolite called pyruvate. The glycolytic pathway consists of a series of substrates, enzymes, and products. Along the way, electrons are collected from the intermediate metabolites and concentrated by electron-capturing molecules called NADH and $FADH_2$. In the beginning of the pathway, some ATP energy is used up; but later on this energy expenditure is rewarded with a net profit of ATP, which can be used for other life-giving biosynthetic purposes. Again, the sugar is split into two parts, each part being metabolized further to make 2 molecules of the pyruvate.

An important biochemical action in glycolysis is called substrate-level phosphorylation. Here, a given substrate that is phosphorylated, i.e., a phosphate attached to the enzymatic substrate, is utilized by a protein-based enzyme to transfer the phosphate to an ADP recipient molecule, making a new ATP molecule—the source of the phosphate for making ATP is a substrate—this was a new type of reaction at the time, and it occurs in glycolysis twice. It is interesting also that in cancer cells, this glycolytic pathway is usually working in a rapid mode, as cancer cells require lots of energy to grow.

Glycolysis is virtually a universal process for all living cells, whether they be bacterial in nature or human. It turns out that because we are carbon-based life forms, the most common source of carbon comes from sugar. Another interesting thing about glycolysis is that it is a central pathway. That is, whether the foods ingested are carbohydrates, protein, fats or nucleic acids, they will all be metabolized via glycolysis.

5) What is the relevance of glycolysis to the Krebs cycle?

In a sense, glycolysis represents an incomplete metabolic system. Its end-product, pyruvate, is not necessarily a complete oxidation—the sugars and other food products can actually be broken down even further, all the way to water and carbon dioxide. The Krebs cycle plays a key role in this complete oxidization. Incidentally, the Krebs cycle has also been called the TCA (tricarboxylic acid) and the citric acid cycles.

In the cases of some intermediary metabolites, they can go straight to the Krebs cycle. On the other hand, for other metabolites, before the Krebs cycle pathway can operate, the glycolysis pathway must first do so. It is in this latter situation that the work of Lipmann comes into play. Glycolysis and the Krebs cycle are connected by a reaction that Lipmann discovered. The Krebs cycle pathway allows the complete oxidation of the metabolites produced in glycolysis.

6) Apparently Krebs and Lipmann shared the Nobel Prize - Can you tell us the nature of their discovery or what they were recognized for?

You are correct about these two amazing scientists. In the year 1953, Fritz Lipmann and Hans Krebs were to share equally the honor of receiving the Nobel Prize in the category of Physiology or Medicine for their discoveries. Prof. Krebs elucidated and connected key biochemical reactions and literally cycled them upon each other to propose the cyclic nature of the pathway, the Krebs cycle. These were largely the contributions of Krebs.

Lipmann, on the other hand, elucidated the inner workings of what is considered to be the most important and central biochemical reaction ever discovered. In short, Lipmann's Nobel discovery identified the entry point into the Krebs cycle and in doing so, made the hugely important connection between the glycolytic and TCA pathways.

Both of these pathways participate in the complete oxidative degradation of sugars, amino acids, fatty acids, glycerol, and nucleic acids while generating CO_2 and water and retrieving their electrons. The electrons are then taken to the respiratory chain so that the O_2 can accept these electrons, generating proton (or other ion) gradients to be used as energy in order to make ATP energy.

In summary, Lipmann's famous reaction, linking glycolysis with Krebs, can be denoted as follows:

$$\text{Pyruvate} + \text{Coenzyme A} + NAD^+ \rightarrow \text{Acetyl-Coenzyme A} + CO_2 + NADH + H^+$$

The enzyme that catalyzes Lipmann's important reaction is called the Pyruvate Dehydrogenase Complex (PDC). The pyruvate was generated as the end-point of glycolysis. The Coenzyme A (CoA) or "Coenzyme for Acetylation," has parts of the B_5 vitamin called pantothenic acid. The NAD^+ is a carrier of electrons and is used as a vessel for the electrons that are taken away from pyruvate; that is, pyruvate oxidation occurs during the reaction. The electrons go to NAD^+ along with hydrogen atoms, called hydrides, which are also taken away from pyruvate to form H^+ (a charged hydrogen atom called a proton) and to make the NADH; the letter H of the NADH stands for the element hydrogen—thus, removal of the electrons (oxidation) from the substrate and accompanied by hydrogen atoms, is a process that is referred to as dehydrogenation. Hence, the enzyme gets part of its name, dehydrogenase, from this step; the hydrogens are lost from pyruvate.

Pyruvate is a 3-carbon molecule, and when a decarboxylation step (defined as a release of CO_2) occurs to form the freed up CO_2, the remaining 2 carbons that are still attached to each other are referred to as an acetate. In Lipmann's reaction, CoA is attached to the 2-carbon acetate, forming Acetyl-CoA. Some investigators call this an acetyl transfer.

Acetyl-CoA happens to be one of the most important metabolites ever made by living organisms. One reason for its relevance is because it's a central metabolite between the two catabolic and anabolic systems that occur during metabolism. Lipmann and colleagues frequently referred to it as an active acetate.

Another reason why Acetyl-CoA is so critical is that it participates in the first and last step(s) of Krebs. Here, the enzyme is called citrate synthase, and like the enzyme says, it synthesizes citrate by combining the 2-carbon acetate with the 4 carbons of the end-point metabolite of Krebs, called oxaloacetate, forming the 6 carbon citrate and completing the cycle!

Thus, applying Lipmann's famous reaction as an entry point into the Krebs cycle was a key finding.

7) I have been to Gamla Stan which is where they award the Nobel Prize in Stockholm, Sweden. For a scientist- is the Nobel Prize basically the highest recognition a scientist can receive?

Indeed, the Nobel is the greatest prize that a scientist could ever hope to garner. Amongst scientists and many others in non-scientific fields of study, there is no other equivalent. The bestowment of the Nobel often signifies general and wide acceptance of the scientific finding, especially amongst a Laureate's peers. Often, when the stakes are high, an important scientific finding may have one or more detractors, and receipt of the Nobel often serves as a confirmation of the validity of these discoveries.

Acceptance of a scientific discovery involves a review of the findings by other experts in the field (often anonymously) prior to the publication in the scientific journal. The quality of the hypothesis, the logical rationale for the hypothesis being tested, the experimental design, the chosen methods of statistical analyses, the repeatability of the findings, and the importance of the discovery are just some of the things that are heavily scrutinized by peer referees or reviewers, before an editor ever accepts the work for permanent publication. Receipt of the Nobel signifies not only a true confirmation and peer acceptance of their discoveries but also of their immense relevance. In short, the Nobel may be the greatest approval of all for an investigator who seeks the truth in scientific terms.

8) What have I neglected to ask about Lipmann?

In his memoirs, there is an interesting anecdote about Lipmann and his attempt to find a job. His post at Cornell not being a permanent job, he obviously needed a permanent post. His golden opportunity arrived in 1940 when Lipmann was to give a seminar about his Copenhagen work at an important scientific conference in Madison Wisconsin. It was a large symposium dedicated to intermediary metabolism. Virtually all of the major experts in the field would be there at the symposium. With all of his experimental accomplishments and discoveries published in the journals, Lipmann was already well known from the literature, and the symposium was his chance to shine in front of his colleagues. It turned out to be the worst presentation Lipmann had ever given.

First, he was not, at the time, terribly experienced in lecturing, and not having prepared properly for the seminar, he ran out of time. He had not anticipated the time constraints being held for each of the talks, and he consequently lectured on only half of the material before he was clocked out. Because of this one disastrous lecture with all or most potential hiring managers in attendance, Lipmann attributes his later difficulties in acquiring a job directly to this one poor performance.

Chapter 24

HANS KREBS

1) Professor Varela, there are some outstanding scientists who made amazing discoveries, and there were other scientists who established a foundation for future work. Now, Hans Krebs actually had a cycle of metabolism named after him. Let's start by defining some words. When you say the word metabolism, what does it mean to you as a scientist? And does it mean something different to the layman, like myself (who is trying to lose weight)?

I have to say that of all of the amazing scientists whom we've been reviewing in this series, Prof. Dr. Sir Hans Krebs is one of my all-time favorites. He was a key figure in establishing the extremely important scientific field known as intermediary metabolism. In fact, it has been argued that Dr. Krebs is the architect of this metabolic domain.

In laymen terms, metabolism often refers to the internal activities inside the human body that regulate body weight and confer a general state of healthiness. Simply put, if one wants to lose weight, one "merely" needs to not overeat during meals, avoid snacking between meals, and (with a physician's approval) exercise regularly, of course.

Scientifically speaking, the term metabolism refers, in general, to the entirety or the sum total of all of the chemical or biochemical reactions that happen to occur in living systems. All living beings on Earth metabolize. There are basically two sub-categories that fall under the overall umbrella of metabolism.

The first metabolic sub-category is called catabolism. Here, foodstuffs are broken down into smaller molecules, some of which are called building blocks, thus releasing the energy stored in the foods. During catabolism, the foods and their smaller chemical intermediates are oxidized, meaning that their electrons are taken away. Frequently, many of the foods are broken down all the way to water and carbon dioxide molecules.

The second metabolic sub-category is called anabolism. Here, small biochemical building blocks are used to synthesize larger molecules, sometimes called macromolecules, that are needed for living beings in order to live. Anabolic reactions often require energy in order to proceed to completion.

All of the stages and the players which function during metabolism that fall in between the foodstuffs eaten by a living being and the many types of the new molecules synthesized, that is, the enzymes, substrates, products, co-factors, energetic mechanisms, etc., constitute the components of intermediary metabolism. There are two types of biological energy; the first is called ATP, which stands for adenosine triphosphate, and the second is called an ion gradient, but is part of, for example, an overall proton motive force.

In summary, when foods are catabolized, the electrons are taken to a so-called final electron acceptor, such as oxygen, or an inorganic molecule or an organic molecule. During the electron transport, protons or sodium ions are pumped across the membrane, building up on one side of the membrane. This motive force is the energy used to make ATP. The ATP in turn is used to make or biosynthesize molecules needed for life to occur.

2) Where was he born and where did he go to school and how did he first get interested in science (do you know?)

Hans Adolf Krebs was born on the 25th of August, 1900, in the then beautiful northern city of Hildesheim, located in Hanover Province, Germany. His parents, George and Alma Krebs, were Jewish. Krebs had two siblings, Elisabeth (5 years older and nicknamed Lise or Lisa) and Wolfgang (2 years younger and nicknamed Wolf). George Krebs was a physician and Alma Davidson Krebs was a stay-at-home mother. Living in a large country house filled with books and harboring an outside garden, the family had many cultural and intellectual pursuits, including reading, theatre, museums, local concerts, and learning to play music—Krebs played the piano.

Krebs attributes his interest in living beings to his father, who taught the young Krebs about the beauty of the countryside and its associated birds, plants and trees. Krebs was an avid reader and had a deep desire to constantly acquire new knowledge, but his parents had encouraged the analysis of such knowledge, as well.

Krebs attended the Gymnasium (grammar school) called Andreanum between 1910 and 1918. Another famous alumnus of the school was Georg Philip Telemann. Although Krebs had an interest in all subjects, his favorite at the time was History—he was, many years later, to write a series of historical works. In his autobiography Krebs notes that both his parents and teachers had plenty of criticism for him, causing him to have the impression of being both unattractive and unpopular. His father's criticisms caused him to be uncertain of his putative potential and abilities.

Yet, Krebs, admiring how his father helped his patients in a positive manner, was influenced to become a physician, too. Just prior to his finishing school in 1918 at the Andreanum, Krebs was drafted into the military, as the Great War (World War I) was in full swing, having started in 1914. The military stint, however, surprisingly lasted only two months, at which time Germany had collapsed, and the war was over. In the newspapers, which Krebs regularly read, the German civilian populace had been made evidently unaware about the extensive nature of the precariousness of Germany's war situation.

Given a fast-track to graduation at the Andreanum because of his military service, Krebs was allowed to take his final exams early. He then entered university at Göttingen in December of 1918. He stayed only one term. It is in Göttingen during a lecture about inorganic chemistry that Krebs was reported to have said to his friend and colleague Erich Stern, "Oh God, I can never learn this." In mid-1919, leaving Göttingen, Krebs moved to Freiburg University, Germany; but shortly after his arrival to the city of Freiburg im Breisgau, his mother died suddenly of influenza. The world was in the midst of the worst flu pandemic it had ever experienced, and an estimated 50 to 100 million had perished during this time.

While at Freiburg, Krebs took courses in bacteriology, anatomy, and physiology, but became interested in research after listening to faculty talk about their various research projects during their lectures. Thus, entering the research laboratory of Prof. Wilhelm von Möllendorff, Krebs conducted studies in the methods of histological (tissue) staining and published his very first paper at 22 years of age—the paper was written by Krebs with very little revision by Möllendorff and immediately submitted to a journal. This was an early demonstration of the scientific acumen possessed by Krebs, as decisively recognized by his research supervisor.

Although Krebs had a very successful start performing research with Dr. Möllendorff, it was, however, Prof. Franz Knoop, and his lectures in 1920 about fatty acid metabolism and its catabolic process of β-oxidation, who first influenced the young Krebs to pursue research in what was to eventually become part of the field of intermediary metabolism. In March of 1921, Krebs passed his pre-clinical or preliminary examination, obtaining the equivalent of the Second M.B., tantamount to a bachelor's degree in medicine.

Continuing at Freiburg for an additional semester, Krebs studied pathology under the tutelage of Prof. Ludwig Aschoff, a teacher whom Krebs considered inspiring. Next, Krebs went to Munich for clinical studies in medicine, spending a brief semester interlude at Berlin, and back in Munich completing his clinical courses of study and passing his final exams in December of 1923.

In January of 1924, in order to attain a medical licensure, Krebs began the requisite year of clinical hospital work at the Third Medical Clinic, associated with the University of Berlin, in Germany. During these pre-clinical and clinical medicine years, post-war Germany suffered from instability, including unchecked inflation. It is also the same time period in which Krebs solidified his interest in pursuing scientific research, instead of practicing medicine. Thus, he entered the laboratory of Prof. Anneliese Wittgenstein and conducted research on the entry of certain anionic dyes through the blood brain barrier and into the cerebrospinal fluid, eventually publishing the work later, in 1926.

Shortly after the start of 1925, in January, Krebs took his M.D. degree and then joined the laboratory of Prof. Peter Rona, housed in Berlin at the Pathological Institute of the Charité Hospital, where he (Krebs) learned techniques useful in biochemistry. Rona's laboratory is famous for having trained other prominent biochemists, such as Ernst Chain, Fritz Lipmann, Karl Meyer, and Hans Weber.

In January of 1926, to his great delight, Krebs moved to the Kaiser Wilhelm Institute for Biology to become a research apprentice in the laboratory of the very famous Dr. Otto Warburg. The Kaiser Wilhelm Institute for Biology is well-known for having housed many famous biochemists, such as Fritz Lipmann, Otto Meyerhof, Lise Meitner, Carl Neuberg, Severo Ochoa, to name only a few.

Warburg's studies in photosynthesis, respiration, and intermediary metabolism were extremely important and influential to the field of biochemistry. Warburg had discovered that tumors metabolize sugars at higher than normal reaction rates even when oxygen is present—so called aerobic glycolysis, becoming part of metabolism called respiration. Hence, the "Warburg effect" became useful for diagnosing and treating cancer.

In Warburg's laboratory, Krebs first studied an enzyme, called cytochrome oxidase, of the respiratory chain, an area for which Warburg would later be awarded the Nobel in Physiology or Medicine, in 1931. Today, the presence of this respiratory enzyme is tested in basic Microbiology Laboratory courses, the oxidase test; and it is often used to help identify the species of bacteria. Importantly for Krebs, he learned the methods associated with measuring metabolic respiration, such as manometry, used in the day to measure oxygen consumption and carbon dioxide production resulting from respiration during the process of intermediary metabolism.

3) Cycles are obviously important things to know about- in philosophy, in history, but why are cycles so critically important in science, and in this case, the cycle of metabolism?

In metabolism, the cycles become important in re-supplying elements like carbon, nitrogen, oxygen, sulfur, and in recycling intermediates that are formed during nutrient metabolism, such as citrate and other central metabolites, back to the living systems, ranging from the bacteria, the plants, animals, to humans and to every living being in between. Without these elemental and nutrient recycling processes in place, these important metabolic molecules would be locked up into unattainable forms, and thus permanently used up. Thus, living organisms would eventually starve and die, and life on Earth could very well cease to exist.

It is during the late 1920s in Warburg's laboratory where Krebs learned the then powerful technique, called manometry, which would help him make his first major contribution to the field of intermediary metabolism—the urea cycle, in 1932. Krebs studied the catabolic degradation of amino acids and the excretion of the nitrogen waste in the form of urea.

This is actually the very first metabolic cycle to have been discovered, and it was Krebs who discovered it. The urea cycle figures prominently in the metabolism of amino acids, which then are used to produce proteins, and which, in turn, carryout life's existence. Hence, the urea cycle is an important topic that is still covered in any good course or textbook dealing with Biochemistry.

There is, however, another cycle for which Krebs was to become famous. Although popularly known as the Krebs cycle, Krebs himself actually referred to it as the citric acid cycle, and others have called it the tricarboxylic acid cycle (TCA). In the strictest sense of the word, the cycle *per se* consists of 10 metabolites and 8 enzymes in many well-studied eukaryotes and with minor variations 11 metabolites and 8 enzymes in prokaryotes. In eukaryotes the Krebs cycle occurs in the inner matrix of the cellular organelle called the mitochondrion. Whereas in prokaryotes, the Krebs cycle occurs in the cytoplasm of the bacterial cell.

During the 1930s, Krebs elucidated his famous nutrient cycle by studying pigeon breast muscle tissue, which was known to carry out metabolism at rapid rates. He added various starting materials to the minced pigeon tissue in the manometer and measured the resulting consumption of oxygen during the ensuing respiration activity inherent in catabolic metabolism. Interestingly, when he artificially supplied the pigeon breast muscle in the manometer with citrate, he observed its depletion and then saw its regeneration! It was then a matter of identifying the intermediates that occurred in between the citrate removal and its reappearance. Adding the requisite intermediates into the manometer along with pigeon muscle, Krebs then showed the re-appearance of the citrate—hence, the cyclic nature of the pathway was established.

It is interesting that Krebs cited the works of Martius and Knoop (his former inspirational teacher) in which they found that oxaloacetate plus pyruvate could make citrate, to argue his (Krebs) case for the cycling aspect of the metabolic pathway.

Unable to identify one of the metabolites in the cycle, however, Krebs simply called it "an unknown substance," in this famous Nobel winning paper of 1937. We now know from the work of Fritz Lipmann, who discovered the central metabolic reaction converting acetyl-coenzyme A (AcCoA or AcSCoA) to citrate by combining it with oxaloacetate that filled in the missing link, so to speak. When glycolysis takes glucose that we eat to form pyruvate, it loses carbon dioxide (which we then breathe out) and the remaining acetate combines with coenzyme A to form AcCoA, which then enters the Krebs cycle to form citrate. The unknown substance that Krebs predicted was the AcCoA. Together, Lipmann and Krebs would share the Nobel Prize in Physiology or Medicine in 1953 for this pivotal discovery.

Interestingly, and importantly, additional work conducted in Warburg's laboratory by Krebs and assisted by a medical student, Kurt Henseleit, the so-called ornithine cycle was discovered, in 1932. Krebs considered himself fortunate to be involved in this discovery for a variety of reasons. First, he chose liver as the tissue to study—it showed high levels of biochemical activity. Second, the ornithine cycle discovery led to another even more important discovery, namely, that of the so-called urea cycle. Third, the discovery of the cycle led to the concept of a metabolic cycle, in general. Fourth, these metabolic cycles were found to be relevant medically and practically in many important diseases. Lastly, Krebs was fortunate that he made the major ornithine discovery prior to the rise in power of Hitler. In this regard, Krebs had officially established himself as a first-rate scientist, and it helped him escape Nazi Germany by accepting an offer from Oxford and later to Sheffield.

Lastly, several metabolic links were discovered between the Krebs cycle and the urea cycle. Collectively called the aspartate-argininosuccinate shunt, the links provided a metabolic conduit between the inner matrix of the mitochondrion, where the Krebs cycle resides, and the cytoplasm, where the urea cycle resides. Anyhow, these two systems are often happily referred to as the "Krebs Bicycle!"

4) What relevance does this cycle of metabolism have today? Is it still discussed?

Among the many justifications for its relevance today, I think there are several primary reasons why the Krebs cycle is still discussed in modern times. The first is that the citric acid cycle is a central repository for metabolic intermediates during the consumption of all foods and synthesis of all important compounds that living organisms need. For example, whether the foods consumed by living beings consist of meat (protein), carbohydrates (sugars), nucleic acids, or fats, all such nutrients are eventually taken to the Krebs cycle; likewise, Krebs cycle intermediates are often the starting points for the bio-production of the amino acids, which are then used for making much needed proteins.

One should keep in mind, then, that there is only so much Krebs cycle machinery present in living cells. Hence, if one constantly snacks or overeats during meals, the catabolic intermediates saturate the Krebs cycle machine, forcing the cell to shunt the AcCoA to produce fats, instead.

Another reason Krebs cycle is important is because it provides a means for living beings to collect and concentrate those delicious and nutritious electrons that are stored in foods we eat in order to take them (electrons) to the respiratory chain to transfer them to their final electron acceptors (oxygen, or perhaps other molecules) to generate an energizing proton motive force in order to have energy to make ATP. Hence, without the Krebs cycle, making biological energy, ATP, might be virtually impossible.

The third reason I think the Krebs cycle is discussed today is that its enzymes can make good targets to inhibit during the growth of tumor tissue, like cancer. The diseases involving cancer remain in modern times a significant cause of mortality.

A fourth reason I think Krebs' cycle is relevant today is because without it there would be no human life—this is not true, however, with many of the microbes, but it certainly is true as far as human life is concerned. With life being such an important entity and, hence, a primary preoccupation, I think it is no wonder that in many scientific textbooks, the cycle of Krebs is quite often given its very own dedicated chapter.

5) What have I neglected to ask about the life and work of this famous scientist?

Altogether, Krebs listed 356 publications in his autobiography, published in 1980, shortly before his death on the 22nd of November in 1981.

Although many biochemists know about this, perhaps it may not be so well known to the layman, but Krebs' well-written manuscript, in which he provided obviously well thought-out and carefully collected manometric evidence for his cycle, was soundly rejected on June 14, 1937, by the editor of *Nature*! It is said that Krebs actually framed the rejection letter and displayed it in his office for the rest of his career. I do know that Krebs faithfully reproduced the rejection letter on page 98 of his autobiography.

It might further interest your readers to know that my postdoctoral supervisor at Harvard, Prof. Dr. Thomas Hastings Wilson, had himself taken his Ph.D. in 1953 in the research laboratory of Krebs, when they were at the University of Sheffield, England. When I was in Dr. Wilson's laboratory during the late 1990s, he was often nostalgic about Krebs and admired him greatly, and I remember thinking how neat it was to know someone who knew the universally famous Krebs.

I think your readers who enjoy music might also appreciate knowing that Krebs happened to come across a song with a description of his cycle as the lyrics, written by biochemists, and which accompanied an old Australian folk song, "*Waltzing Matilda*," but colloquially called instead "*Waltz Round the Cycle*." The story is told that Krebs encouraged these lyricists to keep making more of these scientific songs.

In apparently so doing, there was another song about his urea cycle, titled "*We're Here Because Urea.*" I recall with great fondness that my good friend and colleague, Dr. John A. Badwey, during his lectures on nitrogen metabolism in the large Amphitheatre of the Medical Education Center building at Harvard Medical School, would play this urea tune, to the delight of the postdocs, students, and faculty, alike.

"*Waltz Round the Cycle*"
http://www.csulb.edu/~cohlberg/Songs/krebs.mp3

"*We're Here Because Urea*"
http://www.csulb.edu/~cohlberg/Songs/urea.mp3

Chapter 25

LINUS PAULING

1) Nobel Prize winner (if I am correct) and researcher par excellence, Linus Pauling was one of the giants of the past century. What do we know about his early life?

Linus Carl Pauling, a Laureate of two Nobel awards, was born on the 28th of February in 1901, in Portland, Oregon, in the U.S. He is considered by many to be one of the greatest scientists who has ever lived. His parents were Herman and Lucy Isabelle (nicknamed Belle) Pauling.

Unfortunately, his father, a druggist of modest means, died suddenly of a perforated ulcer in 1910, when Pauling was only nine years old, leaving behind a young widow and three children, two of whom were Pauling's younger sisters, Pauline and Lucile.

During this time, Pauling's mother, Belle, developed a chronic disease known as pernicious anemia, involving a deficiency in vitamin B_{12} and a consequent inability to produce red blood cells. Tragically Pauling's mother was, at the age of 45, committed to an insane asylum, 16 years after her husband's death. A few weeks after being committed, she died of influenza and pneumonia, conditions made to be worse apparently by her chronic pernicious anemia condition.

In his youth, Pauling, a voracious reader and lover of brain teasers, was already considered to be precocious and brilliant. In fact, he was considered, in retrospect, to have been a genius. His interest in the field of chemistry, and later biochemistry, was influenced by a childhood friend, Lloyd Jeffress, who had built a homemade chemistry set. In particular, Pauling had been greatly interested in one of the experiments of the chemistry set, namely, the conversion of sugar to carbon, using a flame. They actually set sugar on fire! Pauling's interest in carbon and later studies of carbon-containing compounds would prove to be of great importance in several of his early scientific discoveries.

When Pauling was a teenager, he came across an old abandoned and defunct metal smelter, with an accompanying old laboratory. Salvaging the old laboratory glassware, equipment, and left-over chemicals from the smelter, Pauling assembled a makeshift but rudimentary and surprisingly sophisticated laboratory, for a youth of that age, presumably in the basement of his house.

2) Where was he educated? (It always amazes me when a group of scholars all come from the same college or university)

Pauling entered Washington High School, in Portland, at the age of 12. Having then taken all of their relevant science courses, Pauling left high school early without having graduated, at the age of 16, in order to enroll at Oregon State University, called Oregon Agricultural College at the time, located in Corvallis, OR. As you indicated in your question, this institution has had many prominent alumni, Pauling among them. Studying chemistry in their main building named Science Hall, but known colloquially as the "Chem Shack" by students and faculty alike, Pauling took his degree in chemical engineering in 1922. During his senior year in college as an instructor of a course in chemistry for economics majors, one of his students, Ava Helen Miller, would later become his wife.

Turning down a graduate fellowship offer from Harvard, Pauling entered graduate school at the California Institute of Technology (Caltech) in Pasadena, California. This is another institution with prominent alumni, currently including as of this writing 34 Nobel laureates. After only three years in graduate school, and having already published 5 scientific papers, Pauling earned his doctorate, *summa cum laude*, in the field of chemistry while minoring in both physics and mathematics, in 1925.

Upon the advice of his Caltech doctoral advisor, Prof. Arthur A. Noyes, the newly minted Dr. Pauling turned down a postdoctoral fellowship offer from Berkeley so that he could accept another offer from the laboratory of Prof. Arnold Sommerfeld, at the prestigious Institute of Theoretical Physics, located in Munich, Germany. This higher educational institution also housed extant scientists and produced alumni who became giants in their respective fields. Pauling's interactions and associations with these numerous investigators facilitated many later productive collaborations.

Just prior to his arrival at Munich in 1926 Pauling received a prestigious Guggenheim fellowship to conduct his postdoctoral research. While in Munich, Pauling was assigned by Sommerfeld the task of calculating the behavior of spinning electrons inherent in a model of the atom developed by Niels Bohr and modified by Sommerfeld—together constituting a so-called Bohr-Sommerfeld model of the atom. As an aside, the story is told that the neophyte Pauling approached Sommerfeld to point out that a paper, previously published by one of Sommerfeld's Ph.D. students, contained an error in its mathematical treatment, which now meant that electron spinning around atoms was not as problematic as previously thought.

Pauling travelled throughout Europe visiting or working briefly in the laboratories of Niels Bohr in Copenhagen, Max Born in Göttingen, and Erwin Schrödinger in Zurich.

During this sojourn, Pauling met many great and soon-to-be-great scientists, including Paul Dirac, Walter Heitler, Fritz London, and Robert Oppenheimer. Shortly after this, Pauling became a faculty member at Caltech, as an assistant professor in 1927 and becoming a full professor of chemistry in 1930.

3) Now, what was his first notable discovery?

While a Ph.D. graduate student at Caltech, Pauling's very first scientific discovery involved the structural determination of molybdenite crystals, publishing this work in 1923. Rapidly acquiring an acumen in the technique of X-ray crystallography, Pauling determined the molecular structures of six additional chemical crystals in graduate school. Despite the presence of so many brilliant and prominent to researchers at such a world class institution, Pauling, nonetheless, stood out as being one of the most brilliant of their students at Caltech.

As a Guggenheim postdoctoral fellow and influenced by the works of Wolfgang Pauli and Erwin Schrödinger, Pauling acquired an expertise in the burgeoning field of quantum wave mechanics. He used his new wave mechanics tools to focus on the behavior of the outermost electrons that orbit larger atoms, and in so doing Pauling made what is considered by many in the field to be his first major and pioneering discovery of note. In greatly simplifying the mathematical approach to wave mechanics as applied to the electron behavior among the large atoms, others could then make significant advances for other elements and then on to even larger molecules. This work was published during 1927 in the prestigious *Proceedings of the Royal Society of London*.

Almost immediately after this publication, Pauling then applied his quantum tools to predict the atomic sizes of ions by examining the behavior of electrons in their outer shells and how they were shielded by electrons in the lower atomic orbitals. These data helped to predict how ions would interact with each other when present in crystals. Publishing this newer work, also in 1927, Pauling solidified his reputation worldwide amongst physicists and chemists as a stellar investigator. Based on these works, others in the field could then combine X-ray crystallography with wave mechanics in order to test new ideas based on each area and to then make new advances in the chemistry of larger molecules, including biologically important ones.

4) What exactly is this alpha helix and what is its relevance to science, biology and chemistry?

The alpha (α) helix is a central structural motif present in proteins of all living organisms. The amino acids that make up protein molecules are arranged like pearls along a string that can form a spiral to make the α-helix structure. Using calculations of bond angles and turning the atoms in the amino acids around each other to make new theoretical hydrogen bonds along the protein chain, the α-helix structure came to Pauling during, the story is told, when he was laid up in bed with a cold—he quite frequently suffered from colds and secondary sinus infections. The theory for the α-helix structure was formulated by Pauling while convalescing.

The α-helix was a key piece of knowledge which investigators astutely used in order to propose the structural model of DNA. Scientists James Watson, Francis Crick, Rosalind Franklin and even Linus Pauling himself applied the α-helix motif of protein chains to formulate the double helix and (inaccurately) the triple helix structural models of DNA.

Students of Biochemistry courses and especially Biochemistry undergraduate majors, are regularly taught all about the famous α-helix structure and its discovery. The hydrogen bonds and the peptide bond angles, their bond rotations, the side chain orientations, etc., in the α-helix are all classically taught in lectures and are unfailingly presented in any good Biochemistry textbook.

The α-helix is one of my favorite molecular structures; in fact, for my Ph.D. thesis, I studied a bacterial antimicrobial efflux pump that shared an evolutionarily conserved amino acid sequence motif; this particular sequence motif had an α-helix structure to it. I think it's an elegant and beautiful structure.

The α-helix structure figures prominently in the inner workings of many proteins, perhaps, for example, by binding to key substrates in enzymes in order to catalyze biochemical reactions or by changing its protein configurations of transporter proteins during the uptake of nutrients by cells or by expelling toxic compounds (like antimicrobials) or waste products from cells. Virtually all biochemical and cellular functions involving proteins will have their molecular mechanisms carried out by protein structures composed of the α-helix.

5) I know that the British used to become ill on long ocean voyages until they began to eat limes, thus becoming known as "limeys," and, of course, limes contain vitamin C. But how did Linus Pauling take this idea and link vitamin C to cancer prevention? (and I have to tell you, I am going to buy some more vitamin C on the way home today!)

The story of limes and of British sailors is an interesting one. Vitamin C, which is also referred to as ascorbic acid or ascorbate, is present in certain fruits, like limes, vegetables, and foods, like sauerkraut. These types of vitamin C-containing foods were indeed often brought along during prolonged sea voyages, in order to prevent a serious disease called scurvy. This condition is brought about by a deficiency in vitamin C.

The discoverer of vitamin C, Prof. Albert Szent-Györgyi, earned the Nobel Prize in Physiology or Medicine in 1937, for this work. Many animals still harbor the four-step biosynthetic pathway to make vitamin C. Such animals don't require dietary vitamin C. In the course of our evolution, however, human beings have lost the fourth enzyme in the pathway that makes the vitamin C molecule. Hence, vitamin C must be supplied in our diets, and daily nutritionally balanced meals usually suffice in filling this bill. The current U.S. recommended daily allowance (RDA) of vitamin C in adult males is 90 mg while that in adult females is 75 mg.

Vitamin C is 'vital' because it plays a role, for instance, in our collagen producing machinery by allowing the amino acid proline to undergo biochemical processing to make 4-hydroxyproline during a reaction catalyzed by the enzyme prolyl hydroxylase, a member of the dioxygenase family of enzymes. Without this particularly helpful vitamin C, a person suffering from scurvy will experience a degeneration of their bones and connective tissues, resulting in the hemorrhaging of their blood vessel, loss of their teeth, slowed healing of their wounds, become prone to gum infections, and suffer heart failure. In mild cases of vitamin C deficiency, patients often experience severe tiredness, irritability, and enhanced frequency of infections in the respiratory tract.

Furthermore, vitamin C has been demonstrated to be an effective anti-oxidant, preventing damage to metabolic building blocks for DNA and protein biosyntheses. The biochemistry of the antioxidant property of vitamin C has been demonstrably worked out. Vitamin C has also been shown to be important for proper functioning of various components within the immune system.

Pauling's interest in vitamin C began in 1965 with what one might call a 'crank letter.' Being a famous person worldwide for making so many discoveries in so many distinct areas of science and, especially, for having won the Nobel in Chemistry in 1954 and the Nobel Peace Prize in 1962, Pauling regularly received many letters of a somewhat dubious nature. The letter in particular was from an industrial brewing biochemist by the name of Irwin Stone. Pauling and Stone had previously met briefly during a ceremony honoring Pauling with the so-called Carl Neuberg Medal, an award for having made a significant contribution linking biological knowledge with that of medicine. In Stone's letter, he reminded Pauling that at the ceremony he (Pauling) remarked that he would like to live 20 years longer in order to witness the new advances occurring in the sciences. Stone is recorded to have asked why only 20 years—why not 50 years longer? Stone went on to say that vitamin C was the key. Stone maintained that just taking enough vitamin C to prevent scurvy was not sufficient for good health. Stone held that mega-doses of the vitamin helped him rapidly recover from severe injuries sustained during a serious car accident.

In retrospect it is easy to see how Pauling was hooked, because he already had an interest and notion of the relative non-toxicity of the vitamins. From a psychiatrist friend, Pauling had learned about, for instance, the efficacy of niacin (a B vitamin) for mental disorders. Additionally, Pauling's good friend, Prof. George Beadle, and 1958 Nobel laureate for his work on metabolic regulation by enzymes that are specified by genes, told Pauling about his work with mutants of his microbe, a fungus called *Neurospora*, which needed vitamins in order to grow.

Thus sufficiently encouraged, Pauling started taking daily mega-doses of vitamin C, on the order of 3 grams, many times the RDA value. According to Pauling, the benefits were immediate and remarkable. Straightaway, Pauling noticed that he had more energy, and, importantly, his frequent colds and their often accompanying sinus infections, a real nuisance for Pauling, had all ceased to occur. Thus, suitably converted, the mega-dose regimen was now part of Pauling's daily routine. Yet, he did not publically advocate vitamin C till three years later, when he suggested its use for mental illness, a treatment he called 'orthomolecular medicine,' in which the most appropriate amounts of a missing substance, a substrate or enzyme co-factor, could be administered to patients, like insulin for diabetes, or fluorine for dental caries.

Pauling's interest in cancer and vitamin C began in 1971 with another letter. This time it was from a physician, Ewan Cameron, who claimed that vitamin C at 10 grams daily was slowing tumor growth in his cancer patients. Since then, the issue of cancer prevention with daily mega-doses of vitamin C has been controversial. Many groups have reported success with cancer and vitamin C, and seemingly just as many others have reported no effect at all, with both proponents and opponents apparently deeply mired in a quagmire. One problem inherent in the issue is that cancer is a multitude of illnesses, depending on, for instance, the nature and extent of the affected cell, tissue, and organ. Also playing important factors are how early diagnoses are made, the level of natural resistance (immunity) of the patient, genetic pre-disposition or hereditary history, and among other factors how healthy a person may be, in general. One has the impression that it will be a very long time before the issue is clearly and finally settled.

6) What were some of his other discoveries and accomplishments?

In a word, Pauling's important discoveries and accomplishments were plentiful. I shall briefly highlight only a few. Let's first address the work for which he earned the Nobel Prize in Chemistry in 1954. The award was bestowed to Pauling for his studies on the so-called nature of the chemical bond and its uses in helping to determine the molecular structures of other important molecules, many of which contained carbon. Pauling had the astute realization that the chemical bonds between carbon and nitrogen atoms, called peptide bonds, were planar (flat) and that their carbon bond angles were tetrahedral (making 4 bonds) in nature—producing a structure appearing like that of the edges of a pyramid.

In addition to discovering the α-helix structure, Pauling also participated in the elucidation of the so-called beta (β)-strand nature of the polypeptide. When folded properly around themselves in a zig-zag type of conformation, the β-strands produced a so-called β-sheet structural motif. These works ultimately led to the elucidation of structures for other biologically important molecules, like hemoglobin and many proteins that are functionally important in the immune system.

Pauling also had an interest in determining the structure of DNA, but unfortunately he lost the 'race' to solve the structure first when he had inaccurately proposed a triple-helix model, instead of the correct double-helix molecule. Francis Crick and James Watson, who had accurately predicted the double helix structure, had also previously and inaccurately proposed a triple helix version themselves. Interestingly, there is the popular story that at the time because of Pauling's political leanings and activists movements, his passport was revoked, and, consequently, he was prevented from attending an important scientific conference in which he would have no doubt heard about the important X-ray crystallography data by Dr. Rosalind Franklin—the same data upon which Crick and Watson had used to produce their accurate double helix structure.

In the story it is speculated that perhaps, if Pauling had only seen Franklin's good X-ray data himself, he may have beat Watson and Crick to the correct double helix DNA structure. While it is true that Pauling most certainly had political enemies, was intensely investigated by Hoover's FBI and that his passport was denied, the other aspects of the story have been nonetheless refuted.

For example, it is said that Pauling's close collaborator Robert Corey did attend the conference and saw Franklin's data. It is further said that Pauling was later able to retrieve his passport and visit the European laboratories at a later time—although it is recorded that Pauling did not visit Franklin's laboratory. Pauling himself later remarked that his push to study DNA was not as much as had been anticipated by Watson and Crick.

Pauling's contributions to the development of the fields of molecular biology and immunology are also significant. While collaborating with another famous scientist, Max Delbrück, early on they predicted that DNA synthesis, a process also called replication, occurred in a complementary way (i.e., complementarity). Similarly, Pauling had predicted that antibodies combined with their corresponding antigen molecules in a complementary fashion. In this way, the molecular structure of the antibody would be matched by the counterpart structure of the antigen in which they would correspondingly fit together in a precise molecular manner.

Pauling also had an interest in the inner workings of the active sites of enzymes, proteins which catalyze chemical reactions in their active sites producing new products but are themselves unchanged during the biochemical reactions. He showed that these enzymatic active sites are stabilized in order to mediate their chemistry activities. Additionally, Pauling formulated the notion that the debilitating disease called sickle-cell anemia could be due to an aberrant hemoglobin structure encoded by a defective gene. Pauling called this situation a molecular disease and started what is now known as molecular medicine within the overall field of molecular genetics.

Pauling and his wife are also noted for their extensive political anti-war and anti-nuclear activities. Their activist efforts where far-reaching and effective in influencing nuclear policy. In this regard, Pauling was awarded the Nobel Peace prize in 1962; he had regretted that Ava did not receive the prize along with him. Pauling is, nevertheless, in the rare situation of having earned two Nobel Prizes, each bestowed to a sole recipient, and each in different categories.

7) What have I neglected to ask?

Of the amazing scientists discussed in this series, Pauling is thus far the first of whom I have had the great opportunity and pleasure to see in person. Furthermore, the very second time I was in the same room with Linus Pauling he had a profound and irrevocable effect upon me.

When I was an undergraduate at the University of New Mexico in the 1980s, I attended an evening seminar given by Dr. Pauling. I remember that the largest auditorium on campus with just under 900 available seats was filled to capacity. After all of these years, all I can remember about the topic that night is that he spoke about a 'three-electron bond' which I had never heard about prior to, or since, then. In any case, even though he didn't mention vitamin C at all, I was amazed, impressed, and deeply grateful that I had had the once-in-a-lifetime chance to hear the famous Prof. Linus Pauling speak.

The next day, in telling a good friend all about the great Linus Pauling and his fantastic evening seminar, I didn't need much convincing to eagerly attend his next technical seminar, scheduled for that afternoon. Thinking we were "sneaking in" to the smaller and seemingly private chemistry seminar held that day to hear Pauling, we happily sat in on this seminar, too. We went early and sat in good seats. We were pleasantly surprised when no one kicked us out, so we stayed. This time the 100 plus seating capacity in the technical seminar room, adjacent to the large auditorium, was overfilled, with many sitting on the floor and people standing at the doorways. The next thing that happened during that seminar given by Pauling changed me forever.

About 15 or 20 minutes into Pauling's presentation, a professor in the audience raised his hand to ask about vitamin C! Now mind you, the actual topic of the technical seminar is lost to memory, but I do know that it was NOT about vitamin C. In his response to the wayward question, Pauling casually mentioned his 'Freshman Chemistry Textbook.' Pauling said that every single thing he wrote in his book, he understood—there was a murmur amongst the professors and students alike in the room. I think that we were all probably astonished that one could understand every single thing in a thick textbook, especially a chemistry textbook! Being an undergraduate, I know I was.

But then he stated that many textbook writers include topics in their books that they themselves don't understand, but yet fully expect their readers to do so—he then added that many professors quite often lecture on topics which they don't understand either, just as much as they lecture on the topics they do understand. It was at that very minute that I promised myself that if I were ever to become a professor that I, too, would teach only that which I understood myself. To this day, this is precisely how I teach.

Chapter 26

JACQUES MONOD, FRANÇOIS JACOB AND ANDRÉ LWOFF

1) Professor Varela, often science and discoveries are the result of true collaboration. It seems that Jacques Monod, François Jacob and André Lwoff had to work together in terms of their discoveries regarding the genetic control of first, enzyme and then virus synthesis. First of all, can you give us a brief summary about these three scientists?

The trio of French scientists to whom you refer, Drs. Jacques Lucien Monod, François Jacob, and André Michael Lwoff, are early pioneers whose bacteriological and bacteriophage genetic experiments were of critical importance to the birth of the molecular biological fields of study. Together, the trio of scientific investigators Monod (a biochemist), Jacob (a biological naturalist and physician), and Lwoff (a microbiologist) received a share of the 1965 Nobel Prize, for their work in the categories of physiology or medicine.

Jacques Monod was born on the 9th day of February in the year 1910 in Paris, France. His parents were Lucien (father) and mother Charlotte (affectionately nicknamed 'Sharlie') MacGregor Todd. Lucien was French, and Sharlie was an American. Accounts report that the Monod household was a center of intellectual and cultural interests, with emphases on the arts, music, and the sciences. When Monod was a young child, the family moved to Cannes, where they lived a largely provincial life. Although Lucien was an artist, Jacques later attributed his early interest in biology directly to his father, who was an early fan of Charles Darwin and of evolution. Monod was enrolled in the College at Cannes (which later became a Lycée) until the year 1928 when he was 18 years of age. He was then enrolled at the Sorbonne, in Paris, where he majored in the field of natural sciences, namely, biology. At the Sorbonne Monod took his undergraduate degree in 1931.

After spending brief periods of study at the University of Strasbourg and University of Paris, in 1932 Monod became a lecturer of zoology. Encouraged by Boris Ephrussi to study genetics, Monod took the advice. He entered the famous research laboratory of Prof. Thomas Hunt Morgan, who was housed at the prestigious California Institute of Technology, where he was taught the fundamentals known in the field of genetics, at that time. Returning to the Sorbonne, Monod became a laboratory assistant in the research lab of Georges Teissier, and it is here where Monod became acquainted with bacteria. Interestingly, it is during this period of the Second World War, in 1940, when Monod had become an incognito member of the French resistance during the German occupation.

In 1941, Monod took his doctorate in the natural sciences. He served as a research professor at the Sorbonne and later at the Pasteur Institute in Paris, where he ultimately became its director in 1971. Monod died of leukemia at the age of 66 years in 1976 on the 31st day of the month of May.

François Jacob was born on the 17th day of June in 1920 in Nancy, France, to parents Simon and Thérèse Franck Jacob, father and mother, respectively. In 1927, he enrolled at Carnot Lycée, located in Paris, where he was a pupil at the institution until 1937. In his autobiography, titled *The Statue Within,* Jacob related how he and his fellow students at the school were brutally traumatized by other students who were right-wing nationalists. A family friend, a surgeon who let the young Jacob observe a surgical operation, inspired Jacob to become a physician. He entered medical school at the Faculty of Paris around 1938. World War II disrupted Jacob's medical education in the month of June of the year 1940 when he escaped to London by boat and joined the military. He was initially stationed in Africa as a medical officer. He was then sent to Normandy where his arm and legs had been wounded severely.

When the war ended, Jacob continued with his medical studies and earned his M.D. degree in 1947. However, due to the nature and severity of his war injuries, Jacob was not able to perform surgery. Thus, he turned his attention to research studies in biology. First, he worked to produce penicillin. Then he was invited by Dr. André Lwoff to join his research laboratory at the Institute of Pasteur, in Paris.

In 1951, Jacob earned a degree in the sciences, and in 1954 he earned his doctorate of philosophy at the University of Paris Sorbonne. In 1956, he became a laboratory director at the Pasteur Institute. In 1960, Jacob became chair of the cell genetics department at the same institution. In 1964, he became professor of cell genetics at the prestigious Collége de France. On the 19th day of April in 2013, Jacob died at the age of 92 years.

André Lwoff was born in 1902 on the 8th day of May in Ainay-le-Château, Allier, France. His parents were Solomon and Marie Siminovitch Lwoff. André's father was a physician who knew the very famous Elie Metchnikoff, discoverer of phagocytes and first immunologist ever to win the Nobel. According to an elegant scientific biography of Lwoff written by his friend and colleague, Dr. Agnes Ullmann, the great Prof. Metchnikoff introduced a 13-year old Lwoff to the microscope. Lwoff was an undergraduate at the University of Paris, where he took his college degree in natural sciences in 1921.

He had been an apprentice of sorts under the tutelage of another great scientist, Edouard Chatton, a protozoological taxonomist. At 20 years of age, Lwoff joined the laboratory of Felix Mesnil (a colleague of Dr. Pasteur) at the Pasteur Institute, in Paris, France, with the encouragement of Chatton. In 1927, Lwoff garnered an M.D. degree. Dr. Lwoff then married Marguerite Bourdaleix, who would become an established investigator in her own right but unfortunately lived under the shadow of her husband.

In 1932, Lwoff earned his Ph.D. at the University of Paris. As a postdoctoral fellow at the University of Cambridge, the Rockefeller Foundation funded his work in the laboratory of David Keilin. Next, Lwoff joined the Pasteur Institute in 1938. In 1959, he became professor of microbiology at the Sorbonne. In 1968, he retired from the Pasteur Institute and took on the directorship of the Cancer Research Institute, in Villejuif, France. In 1994, at the age of 92, he passed away on the 30th day of September.

2) Now, what exactly do we mean by virus synthesis?

Depending on your point of view, a mature virus, called a virion, may be considered amongst the tiniest of all living organisms or one of the most complex aggregates of non-living chemicals. Thus, if you hold that viruses are alive, then you might think that they are *bona fide* microbes who happen to need a lot of help from host cells to grow. On the other hand, if you hold that viruses are non-living, then you might think that viruses are simply virus particles. In any case, a virus consists of some sort of nucleic acid genome, like RNA or DNA, which is surround by a protein coat, called a capsid. Some of the viruses have an additional biological membrane, called an envelope.

A key property of viruses is that they lack biological machinery to make more of themselves. Thus, in order to propagate their numbers, viruses must gain access to the insides of cells, where biological machinery for nucleic acid and protein biosynthesis resides. In a technical sense, viral synthesis today refers to the stealing and using of internal cellular biological machinery in order to synthesize more viruses. We refer to biosynthesis as one of the steps in the infection process.

I think that prior to the 1950s, viral synthesis referred to the propagation, replication, or multiplication of viruses that occur during infection of a host cell by a virus. For our purpose of discussion here, there are two main types of viral infection to consider. Lwoff discovered one of these types of phage virus synthesis.

In general, virus synthesis occurring during infection of a host cell by a virion involves the following stages. **(a)** The virus binding to a host cell receptor; **(b)** entry of the virus or its genome into the cell; **(c)** biosynthesis, involving nucleic acid genome replication, transcription to make RNA, and protein synthesis; **(d)** assembly of viral capsid and packaging of genome into the capsid; and **(e)** escape of amplified virus numbers from the host cell.

The trio Lwoff, Monod and Jacob studied a special type of virus that infects bacterial host cells. These viruses are called bacteriophages or simply phages. Frederick Twort discovered the phages in 1915.

The phages and their host cells were quite easy to manipulate in the laboratory and served very well for the study of genetics. Although there are many types of viruses, I shall consider here primarily the phages.

The first type of viral infection was already known and has been called the lytic cycle, in which the host cell dies by being lysed—this cell lysis releases the multiplied viruses from the internally encased host cell. It allows the viruses to move about in order to seek new host cells to start the infection process all over again.

In the second case of viral infection, discovered by Dr. Lwoff, is called lysogeny. In this case, the infection starts out similarly to that of the lytic cycle, but the outcome of the infection takes a turn that differs from the stages outlined above. Lysogeny involves the following stages: **(a)** viral binding to the host cell; **(b)** viral or nucleic acid entry; **(c)** integration of viral nucleic acid into the host cell genome—at this point the virus is referred to as a prophage, and the bacterial host is called a lysogen; **(d)** binary fission growth of the bacterial lysogen—here, the lysogen can grow indefinitely if given the chance; **(e)** induction of the lytic cycle—Lwoff discovered this by exposing the lysogen to UV light; **(f)** biosynthesis; **(g)** assembly & packaging; and **(h)** phage escape.

3) I have repeatedly heard about *E. coli*- what exactly is it and why is it relevant to these scientists?

The term *E. coli* is the shortened version of the well-known bacterium, which, in its entirety, is officially called *Escherichia coli*. Dr. Theodore Escherich, who had originally named the clinical isolate as *Bacillus coli*, discovered it. It was later named *E. coli* in Dr. Escherich's honor.

In general, there are two main categories of *E. coli* bacteria. The first kind is the pathogenic type, represented by a series of insidious strains of *E. coli*, all of which are causative agents of infectious disease, and some of which are quite severe.

The second type of *E. coli* is the non-pathogenic kind that is typically found to reside in the gut of all animals, including those of all humans. This relatively harmless kind of *E. coli* is often used as an indicator of fecal contamination in environmental locations outside of the research laboratory. Inside the research laboratory, however, it has been a different situation altogether. As a proverbial 'laboratory rat' of sorts, *E. coli* has been used for basic research, such as that seen in both early and now modern molecular biological research investigations.

In their Nobel Prize discoveries, the investigators Monod, Jacob, Lwoff and many others used *E. coli* as the host cells to study the regulation of gene expression, for the first time in history. They demonstrated the presence and activity of the so-called *lac* system, which they later called the *lac* operon.

4) Monod is apparently believed to be one of the founders of molecular biology. If so, where do the other two scientists fit in?

I think that both Drs. Lwoff and Jacob can be considered, with a very large degree of certainty, as *bona fide* co-founders of molecular biology, and on par with Dr. Monod.

As I mentioned above, Dr. Lwoff's work was critical to the discovery of lysogeny. Starting in 1949, he had studied the phenomenon from the perspective of bacterial immunity to phage infection. Lwoff and his co-workers Niels Kjeldgaard and Louis Siminovitch worked with a bacterium called *Bacillus megaterium* because of its large size—which they could exploit to examine a single bacterial cell in the test tube and under a microscope.

Adding bacteriophage to his single cell, they could observe cell lysis resulting from the lytic cycle of phage infection. However, Lwoff and colleagues also observed a great many individual cells not undergoing the lysis and instead remaining intact. These lysogens stubbornly refused to be lysed. Lwoff's team tried every type of agent and failed to observe the lysis, until they tried, of all things, the UV light!

That is, UV light induced the lysogen to undergo the lytic phase of infection. Lwoff was to reminisce in later years that this finding was one of his greatest thrills of his investigative career in science.

Of the many discoveries made by Dr. Jacob, he had two quite notable ones. The first one dealt with following up on Lwoff's findings regarding lysogeny. Jacob and collaborator Elie Wollman focused on bacterial conjugation as a tool with which to study the lysogenic phenomenon using a bacteriophage virus called phage lambda (λ). They found that in the prophage stage of the phage λ, the viral genome integrated into the bacterial host genome, a phenomenon they called an episome (an integrated λ genome). The prophage episome formation occurred during bacterial conjugation.

The second key discovery involving Dr. Jacob involved his work with Dr. Monod and Dr. Arthur Pardee in which they discovered the famous *lac* operon. It was the first time in scientific history that the operon notion was examined and in such genetic detail. They had been interested in distinguishing between two seemingly contradictory hypotheses. On the one hand, they wanted to know whether *E. coli* bacteria underwent induction of the enzymes that were needed to utilize the milk sugar lactose for energy and growth. On the other hand, they also wanted to know how this so-called enzyme induction system related to the outwardly conflicting notion that end-products of enzymes inhibited the production of the enzymes that made the products in the first place, a situation known as repression. In short, they wanted to reconcile the two hypotheses of enzyme induction versus enzyme repression with one grand unifying hypothesis. In so doing, they discovered the *lac* operon system of gene regulation.

5) Monod and Jacob apparently first garnered fame and their reputation for their research on *E.coli lac* operon. What did this research show and why was it important?

The initial principle of the *lac* operon system first came to Jacob while watching a Western film with his wife Lise, during a vacation in August of 1958.

Jacob had thoughtfully compared his studies on the λ phage infection of *B. megaterium* to produce lysogeny with the studies pertaining to the utilization in *E. coli* of the sugar lactose for energy production and growth. Monod had just found *E. coli* mutants that had lost their ability to be induced to make lactose-utilizing machinery—that is, the mutants had lost inducibility and were thus referred to as i^- for loss of inducibility. These so-called i^- mutants (or $lacI^-$) made plenty of β-galactosidase enzyme, a genetic type called z^+ (or $lacZ^+$).

Jacob reasoned that the two systems, i.e., the λ-infected lysogen induction to commence the lytic cycle with UV light and the isolation of the so-called constitutive i^- mutants of *E. coli*, which showed loss of inducibility by continually producing the lactose metabolizing enzyme β-galactosidase, may be operating in much the same way. That is, the UV light may have broken a λ repressor to produce λ phage, and the mutation in the *lac* system may have involved a broken lactose repressor to produce β-galactosidase. In essence, both seemingly different systems, i.e., the λ phage synthesis and the lactose utilization machinery, had a common type of molecular regulatory player: the repressor!

The story is told by Dr. Benno Müller-Hill in his delightful little book '*The* lac *Operon*' that when Jacob first told Monod about this new insight (which was later demonstrated to be correct), Monod at first laughed and then ridiculed Jacob, explaining at length how it couldn't possibly be correct. The next day Monod was converted.

The development of the bacterial conjugation tool (that I mentioned above) by Jacob and Wollman, could be used by Jacob, Monod and Arthur Pardee to map the genomic locations of the genetic mutations by following the time needed for one donor cell to insert its genetic material into the recipient cell. At any point during the 100 minute conjugation period, the DNA transfer could be stopped simply by using a Waring blender (essentially of the same type used in any kitchen) to break apart the conjugating bacteria. In this way, Monod and Jacob could determine how closely linked the gene encoding the repressor protein (denoted LacI) was to the gene encoding the β-galactosidase enzyme (denoted LacZ). It was found that the *lacI* and *lacZ* genes were closely linked genetically. Using the conjugation method, it was also found that the y^+ gene (*lacY*) encoding the lactose permease protein, LacY, was also closely linked to the *lacZ* gene.

The next experiment they did is well-known amongst molecular biologists and a widely celebrated one, which in turn led the way to the full discovery and elucidation of the very famous *lac* operon. This investigative project has been affectionately referred to as the so-called 'PaJaMo' experiment (or sometimes 'PaJaMa' but pronounced pajama or 'pyama'). The PaJaMo experiment was named as such after the first-two letters in the names of its experimenters, **Pa**rdee, **Ja**cob and **Mo**nod.

The PaJaMo experiment involved conjugating a cell with a *lacZ*$^+$ property (makes β-galactosidase enzyme) to transfer its z^+ gene into a recipient cell that was both *lacZ*$^-$ (could not produce β-galactosidase) and *lacI*$^-$ (could make β-galactosidase because its repressor was broken). The donor cell had a constitutive phenotype (made β-galactosidase with or without inducer) while the recipient had an inducible phenotype (made β-galactosidase only with inducer). The conjugated cells made β-galactosidase regardless of whether inducer was present. Thus, the mated bacteria switched their biochemical behavior from a constitutive state to an inducible one!

Taking into account the rates of the β-galactosidase synthesis in the mated bacteria compared to the original parental generations, the concept of a repressor was formulated in which it (the repressor) could be induced to allow production of the β-galactosidase when in the presence of inducer. Monod, Jacob and Pardee reasoned that the DNA element to which the repressor acted upon was called the operator, or *lacO*. Further mapping showed that the genes were arranged as follows: *lacI*, *lacO*, *lacZ*, *lacY*. Thus, the nature of the operon was established.

Further genetic and biochemical analyses showed that the *lac* operon worked in two distinctive modes: repressed versus induced.

The repressed mode was in operation when cells were grown without inducer called allo-lactose, a derivative of lactose. In this scenario, the continually made (i.e., constitutive) LacI repressor tightly binds to the operator DNA, preventing RNA polymerase from transcribing the *lacZ* and *lacY* genes to make RNA and thus preventing the synthesis of β-galactosidase and permease proteins.

In the induced mode, inducer (allo-lactose) is present and binds to the LacI repressor. The inducer-repressor complex is unable to bind the operator. Thus, in this scenario, RNA polymerase binds its promoter, turning on gene expression in the form of transcription and translation of LacZ and LacY.

With the notion of repression versus induction established with the famous lac operon, it paved the way for molecular biologists and biochemists to manipulate the gene expression programs for a variety of other genes. Genes could be turned on or off, as desired. Such gene expression systems could be used for improvements in biomedicine. It led the way to the eventual development of gene therapies, production of important medicines, invention of critical diagnostic tools, and many other important applications.

6) François Jacob made some additional contributions in the realm of genetics by successfully working with other famous scientists like Monod and Lwoff while also working at the famous 'Pasteur Institute' of France. What were some of these contributions?

While at the Institute of Pasteur, François Jacob was key to several important discoveries. He discovered many of the intricate mechanisms involved in lysogeny. He showed the inner workings of prophages and how they integrated as episomes into the chromosomes of host bacterial cells. Jacob helped to develop the bacterial conjugation method for the purpose of mapping mutated genes and studying their resultant behaviors within new host bacteria. His work led to our understanding of gene transfer between bacteria—a key feature now recognized in transfer of antibiotic resistance genes between different species of bacteria.

Jacob was a significant figure in the discovery of key mechanisms in the regulation of gene expression, using molecular players, like RNA as messengers, genes that are regulated, and genetic elements that group together several genes to mediate a particular behavior—that is, the notion of an operon. Lastly, while at the Pasteur Institute, Jacob also participated in the discovery of allostery, a phenomenon in which an effector acts upon an enzyme by binding to a site, which is not its active biochemical site, on the enzyme to alter its structure, resulting in its altered activity. Enzymes are often inhibited in this allosteric manner.

7) André Lwoff in 1954 began studying poliovirus. His experiments on the relationship between the temperature sensitivity of viral development and neuro-virulence prompted him to evaluate the problem of viral infection. How did these investigations lead him to work with Monod and Jacob and how are they relevant?

When Dr. Lwoff began studying poliovirus he measured their virulence capabilities in neuro-tissue after repeated passaging in cultured host cells under varying degrees of temperature. Together with his wife Marguerite Bourdaleix Lwoff (a prominent research investigator in her own right), they found that the poliovirus managed to mutate into highly virulent forms, regardless of the temperatures used during the viral passaging in cells. Whether cold or hot, the new poliovirus variants demonstrated their profound virulence.

Analyzing the various conditions for mutant isolation, whether virological (such as bacteriophages or poliovirus) or bacteriological (such as lysogenic), all were shown to be useful in Lwoff's collaborations with Monod and Jacob. Such work led ultimately to the discovery of the mechanisms involved in the lytic versus the lysogeny cycles of phages and to the innate immune response factors directed against poliovirus.

8) I would be remiss if I did not mention Lwoff's work on bacteriophages, microbiota and on the poliovirus. First, can you explain what each of these are, and how they are relevant to medicine and perhaps to the Nobel Prize he won with Monod and Jacob?

Let us briefly consider first the bacteriophages. These agents are viruses that specifically infect bacteria. As mentioned above, Lwoff used phage λ to solve the mystery of bacterial immunity to phage infection. In so doing, he began the field devoted to the regulation of gene expression using the lambda repressor. Its molecular mechanism was elegantly told in a wonderful book called '*The Genetic Switch*' by Mark Ptashne. As I mentioned above, Lwoff's work was used to help Monod and Jacob to work on the famous *lac* operon. One application of bacteriophages involves an old idea but has recently been somewhat in vogue, and it is called phage therapy. The concept of phage therapy consists of using specific phages to treat bacterial infections. Two recent developments have been published in which specific phages target listeria (*Listeria monocytogenes*) or the MRSA (methicillin resistant *Staphylococcus aureus*) bacteria.

Regarding the microbiota, it refers to the nature and composition of the microbes that reside in a particular environment, such as water, soil, or other organisms. Nowadays, modern microbiologists use the term microbiome to denote the species of microbes harbored in various environments. Lwoff's work in this area was conducted in the 1930s when he had worked under Prof. Edouard Chatton at the Pasteur Institute and when Lwoff had had the Rockefeller grant to work in the laboratory of David Keilin at Cambridge. Under these venues, Lwoff studied various eukaryotic microorganisms such as the protozoan ciliates. The human microbiome today has taken on monumental importance as being involved in many health-related issues, ranging from heart disease, diabetes, and cancer, to human emotion, such as happiness or moodiness.

It may interest you and our readers to know that poliovirus is very likely to go down in history as being the second microbe to be completely eradicated from the planet—smallpox virus was the first. Lwoff had had a hand in starting the polio eradication movement by studying how the innate immune system responds to poliovirus. When Lwoff showed that attenuated poliovirus mutants were temperature sensitive, Albert Sabin was able to exploit this knowledge to develop a viral marker for ensuring quality control during vaccine production. Apparently, the marker test is still used to this day to test live oral poliovirus vaccine preparations.

9) What have I neglected to ask about this group of scientists and their collaborative effort?

I think it is fascinating that the collaborative works of Drs. Monod, Jacob, and Lwoff led to the establishment of many other fields of study within the domains of molecular biology, biochemistry, microbiology, and biomedical science. A brief listing is conveyed here.

First, their studies of microbial adaptation to sugar nutrients and the resulting microbial mobilization of the machinery required to metabolize the various sugars was seminal in the establishment of the microbial physiological field of study. Their study of enzymes like the LacZ β-galactosidase led to further work in protein chemistry, enzyme kinetics, and reporters for gene cloning and protein structure elucidation. Students of biochemistry and even microbiology laboratory courses may often still study the activities of LacZ.

The work dealing with the LacY permease led to groundbreaking studies of sugar transport across the membrane, transporter structure elucidation, energetic mechanisms of solute translocation, and substrate specificities. Incidentally, the trio had mistakenly predicted that the LacY permease was an enzyme, hence, the name permease. The LacY protein is actually a transporter, rather than an enzyme, but the historical name 'permease' has stubbornly refused to go away, despite many attempts to correct it in the scientific literature.

The studies dealing with the LacI repressor have led to astonishingly significant advancements in protein homology structures, protein-DNA binding interactions, regulation of gene expression systems, gene cloning, and in gene therapy. This is a large field of scientific inquiry today involving many prominent investigators.

Their work addressing microbial genetics led to further work on gene mapping, bacterial and viral mutant isolation and mutant characterization, bacterial conjugation, and phage mechanisms of infection. In modern textbooks of biochemistry, cell biology, genetics, molecular biology, molecular medicine, virology, and microbiology, the *lac* operon and lysogeny are often given due prominence.

Chapter 27

ALBERT SABIN

1) Professor Varela, anyone who knows anything about polio, probably knows the name Albert Sabin. First of all, where was he born where did he study and what is he most known for?

Dr. Albert Bruce Sabin is best known for having developed the live attenuated oral polio vaccine (OPV), which was used globally to invoke prevention of polio, a debilitating disease that in its worst form has inflicted suffering for at least two millennia.

Dr. Sabin was born on the 26th day of August in the year 1906 to parents Jacob Saperstein (father) and Tillie Krugman (mother) in Bialystok, of the Russian Empire, in what is now modern Poland.

The Saperstein family changed their name to Sabin after they moved to the United States as immigrants. Sabin enrolled in Patterson High School, in New Jersey, as a sophomore. After HS graduation, Sabin moved to New York University (NYU) where he studied pre-dentistry and spent two years in dental school.

Sabin had been influenced to appreciate the science of microbiology after having read the classic book titled '*Microbe Hunters*' by Paul de Kruif. Furthermore, Sabin apparently had hated dental school, and he equally hated even the thought of becoming a dentist. Therefore, he quit the school, and his prime benefactor, an uncle, cut-off Sabin's primary means of financial support.

Poverty as a college graduate made young Sabin go to work as a laboratory worker under a great mentor, William Hallock Park. Sabin had known Prof. Park when he took his graduate-level bacteriology course back in dental school. Park had been a tremendously helpful mentor to young Sabin, providing him with various opportunities that proved to be quite helpful in facilitating Sabin's later career.

Sabin entered medical school at NYU, while he worked part-time as a laboratory assistant in Park's laboratory, finding little in the way of time for his medical studies. Yet, in the laboratory, Sabin managed to discover a new technique for typing the pneumococci bacteria (nowadays called *Streptococcus pneumoniae*), a method that became known as the Sabin Method.

Sabin first encountered the disease polio from a scientific standpoint while working as a young physician at Bellevue's teaching hospital associated with NYU, when an outbreak occurred in New York City shortly after he took his M.D. degree, in 1931. Sabin had participated in performing autopsies on victims of polio. This experience was later to be of tremendous benefit to Sabin in his work with his polio vaccine.

2) Salk and Sabin are probably the two most famous names in polio research. Did they ever work together?

Jonas Salk and Albert Sabin did not collaborate much with each other. They worked almost entirely independently from each other, and each formulated their own particular type of polio vaccine.

Salk prepared a so-called inactivated formalin-killed poliovirus vaccine, called the IPV. Sabin, on the other hand, prepared a so-called live attenuated oral polio vaccine, called the OPV.

Salk and Sabin were, however, staunch rivals and adversaries. They had known each other personally, speaking with each other at scientific conferences, shared strongly worded letters to each other, and quite often harshly criticized each other's vaccine. Their relationship was contentious; the conflicts between them lasted a full 40 years. They attacked each other personally, and each had allies on their sides, picking up the mantle and continuing the conflict. Their conflict outlasted both Salk and Sabin. Some historians of science maintain that the conflict is in full force to the present day.

3) Sabin also developed vaccines against encephalitis- First, can you please define encephalitis and its symptoms and then tell us how this came about?

Encephalitis occurs when the brain gets infected with a microbe, such as, e.g., a bacterium, a virus, or a protozoan, and causes an inflammatory response in the brain. The condition sometimes occurs as a complication after suffering a bout of meningitis. While the signs and symptoms of encephalitis may vary, common serious ones include blindness, deafness, altered personalities, and coma. The condition may be fatal.

Sabin studied the particular form of encephalitis caused by the so-called Japanese B virus, a known causative agent of the inflammatory condition that is transmitted by insects. While a medical officer during World War II, Colonel Sabin prepared a so-called mouse brain vaccine against the Japanese B virus, a causative agent of the encephalitis condition. While this encephalitis vaccine was administered to thousands of GIs in the Pacific theatre, its preventive efficacy was apparently never fully evaluated.

4) Sabin apparently worked with the Russians as a true humanitarian. What do we know about his endeavors in the Soviet Union?

Sabin's polio vaccine was tested on millions of children, and later adults, in the Soviet Union during the late 1950s. Despite limitations in language, significant bureaucracy, and the height of the cold war in full swing, Sabin and Soviet scientists managed to form a successfully cooperative and organized collaboration. While the Salk vaccine had been tested earlier and on a limited basis in the Soviet Union, it nonetheless showed somewhat 'mixed results' there. Enthusiasm for the Salk vaccine in the Soviet Union diminished, especially because it required a cumbersome and an invasive mode of administration (multiple injections by needles), and it was more expensive to produce than the Sabin vaccine.

When asked to participate more fully with his vaccine in the Soviet Union, Salk refused—he was later reported to have regretted his decision to do so. Thus, Sabin was in place to lobby for his vaccine in the Soviet Union. With the Soviet police state firmly in place, Russian parents were effectively made to have their children efficiently brought to vaccination centers. Russian compliance with vaccination by the OPV was all but guaranteed. There weren't even any non-vaccinated control groups as cohorts, as had been dutifully performed in the US with the Salk vaccine testing, earlier in the 1950s. In the Soviet Union, the prime objective was polio eradication, rather than vaccine efficacy testing.

After a suspiciously exaggerated claim to astonishing success with Sabin's vaccine in the Soviet Union, an independent verification had demonstrated the Sabin vaccine efficacy. A polio epidemic in the Soviet Union had been fundamentally thwarted. Hence, the Sabin polio vaccination effort in the Soviet Union was deemed a largely humanitarian effort.

5) Sabin further traveled to Cuba, Israel and a number of other countries- thus he truly was a world-class scientist and researcher- Your thoughts?

Indeed, with the success of his vaccine in the Soviet Union behind him, Sabin turned his attention to Cuba, trying to establish a similar mode of cooperation between the US and Cuba through their respective scientific national academies. The effort failed.

Later historians, however, speculate that despite this initial failure to establish scientific ties between the US and Cuba, later research ties in the biomedical sciences emerged between the two countries. It is, thus, further believed that Sabin played a role in starting this more recent scientific cooperativity between the two nations.

In Israel, Sabin became president of their prestigious Weizmann Institute of Science around 1969 but had to resign a few short years later, due to a heart condition that required invasive surgery. In later years, in the 1980s, Sabin traveled to Brazil to address a polio outbreak there.

Regarding Sabin's travels abroad, he had countless opportunities to meet with various scientists around the world. Being so universally travelled and globally experienced, I think it has most certainly made him a world-class scientist as you have so aptly put it. Being able to have learned about scientific endeavors in a variety of other cultures had provided Sabin with a widely recognized acumen as a dedicated scientist. His scientific peers certainly recognized Sabin as such, rewarding him with election to the National Academy of Sciences in the early 1950s and with many other scientifically based awards. It seems that with diversity, advantages are enjoyed with having a larger repertoire of tools from which one may realize better solutions to problems, and Sabin certainly had such an advantage, from his world-class status.

6) Quick question- has polio been eradicated due to the efforts of Salk and Sabin?

As of this writing, the short answer to your question regarding global polio eradication is: not yet, but possibly soon, barring no unanticipated glitches.

It is anticipated that polio may be eradicated shortly after this book is published, and both Salk and Sabin will definitely have had something to do with its demise. The worldwide eradication of polio, when it occurs, will have been due to the massive undertaking of vaccination programs, beginning in the early to mid-1950s and with a big push in the late1980s. Salk and Sabin independently developed anti-polio vaccines, each with their distinctive types, origins and modes of operation. Together, their polio vaccines have been used on a massive scale to vaccinate countless millions of children and adults throughout the second half of the 20th century and in the early 21st century.

Eradication of polio from the planet will constitute the second case in recorded history in which a disease has been eliminated as a direct result of human intervention. The first such case occurred in the late 1970s with the obliteration of the dreaded small pox infections, after a concerted effort was instituted in the middle of the 20th century. A third infectious agent is slated to become globally eradicated, barring no unforeseen barriers, such as war or climate. This parasitic worm has the scientific name *Dracunculus medinensis*, but it is known more commonly as the Guinea worm. This infectious worm agent causes an ailment known as dracunculiasis Perhaps the *D. medinensis* agent will also be purged from the Earth soon. As of this writing, the dracunculiasis disease has been limited to a few small African countries.

Regarding the respective roles of the Salk vaccine versus the Sabin vaccine in the polio eradication efforts, both seemed to have played roles depending on the particular eras and geographical locations. During the 1950s the Salk vaccine played a prominent role in stopping the polio epidemic then raging in the US. During the 1960s, however, the Sabin vaccine took over as the predominating polio vaccine that was used in the US. The late 1970s saw the final elimination of naturally produced polio from the US.

While both vaccines had their advantages and disadvantages, one drawback of the easily-administered oral Sabin vaccine was that in very rare instances, poliomyelitis was caused by the vaccine, by inadvertent exposure to others not yet old enough to be vaccinated. At some point during the mid-1990s, clinicians tried using a combination of both types of vaccines—the Salk and Sabin types—hoping to eliminate the rare Sabin live-virus vaccine-induced polio cases. The vaccine combination failed to work toward this effort—the rare vaccine induced cases still seemed to occur, albeit sporadically. Finally, the US switched back to using the killed Salk vaccine in the year 2000 while using the Sabin vaccine for cases of children arriving to the US from endemic countries.

On a global scale, however, the predominating vaccine is presently that of Sabin's.

7) Unlike some other scientists, apparently Sabin did not patent his discoveries and thus did not make very much money from his vaccine. He was indeed a true philanthropist in addition to being a great scientist. Any insights into this?

Sabin's refusal to patent and thus profit from his vaccine had historical ramifications. Sabin managed to outmaneuver his competitors. Except for Salk, most of Sabin's competitors had also developed live attenuated polio vaccines. These challengers had definitively planned to patent and widely market their vaccines for profit. Sabin was quicker to conduct his vaccine testing trials and demonstrate protective efficacy of his vaccine preparation, compared with his competitors. The other competing vaccines were less efficacious in polio prevention than Sabin's.

Taken together with the ease of administration and having developed a trivalent polio vaccine, Sabin's preparation won the global race in the end. Had he not, polio vaccination may have ended up being costly and prohibitive in countries that need it the most, thereby possibly derailing the track to global polio eradication. Fortunately, for the world, such a potentially harmful derailment did not occur.

8) What have I neglected to ask?

Interestingly, both Salk and Sabin vigorously decried the massive bureaucracy that they each encountered during their respective works. Both vaccine inventors, at various times, were stymied in their efforts by a seemingly endless multitude of bureaucrats, e.g., grant evaluators, benefactors, administrators, committees, governments, the Cold War, etc. Sabin had even once claimed that although there was an effective vaccine, it and the scientific process had been hijacked by such bureaucrats.

It seems that neither Sabin nor Salk ever got the Nobel, even though their efforts ultimately have led to the substantial success with polio reduction on a worldwide scale. Each polio vaccine discoverer was criticized variously, but especially in not having made original or groundbreaking discoveries.

Polio research was, however, Nobel worthy, not for Drs. Sabin and Salk, but for Drs. John Enders, Frederick Robbins, and Thomas Weller. These three investigators took the Nobel in 1954 in the area of physiology or medicine. They had been able to culture the microbial causative agent of polio, namely, the poliovirus, in the laboratory within a variety of cultured host cell types. This work subsequently allowed Salk to develop his killed polio vaccine.

Chapter 28

ALFRED HERSHEY AND MARTHA CHASE

1) Alfred Hershey and Martha Chase worked to confirm that DNA is genetic material. Why is this such a breakthrough?

The experimental work of Drs. Martha Chase and Alfred Hershey represents one of the most important scientific breakthroughs in the 20th century. Their work reinforced the notion, more like a hint, that DNA was the hereditary material for passing along newly made genes into later generations of living organisms.

The experimental results of Chase and Hershey helped key investigators to focus their efforts on DNA as constituting the genetic material. Before their influential work was revealed to the scientific community, most of the investigators with an interest in genetics and heredity had firmly held convictions that protein was the prime candidate for conferring heredity. At the time, up until the early 1950s, the incorrect idea that protein was the main genetic substance actually seemed plausible because of the complexity of protein.

For example, protein is made up of 20 distinctive amino acid molecules, whereas DNA merely has a phosphate, a sugar, and one of four distinct nitrogenous bases, called adenine, cytosine, guanine, and thymine. As a result, it was thought that DNA was too simple of a substance to confer heredity upon organisms, with all of their complexities.

To muddle the issue even further at the time, protein and DNA were found to come together during the early DNA purification attempts. These purified substances were known as nucleoprotein, and it was thought (again incorrectly) that the protein part of the nucleoprotein was the heredity material.

Shortly after the Chase-Hershey data became known, about nine months later, James Watson and Francis Crick developed their now famous double helical model for the structure of DNA. In both of their personal memoirs, both Watson and Crick were later to directly attribute the Chase-Hershey results to their main push to elucidate the DNA structure.

2) Watson and Crick seemed to follow in the footsteps of Hershey and Chase with their work. What did THEY find?

In April of the year 1953, Crick and Watson published their famous paper in *Nature* revealing to the rest of the scientific world what the molecular structure of DNA was like. It was Earth shattering—at least it was so to the experts in the fields of genetics, molecular biology, biochemistry, cell biology, and even physics.

Watson had been made acutely aware of all of the intricate workings of the Chase-Hershey experiment. As Hershey was unable to attend an important scientific symposium on the topic of genetics, the task of reading the work fell upon Watson, who basically stood up at the meeting and read the Chase-Hershey paper to those in attendance. Watson was later to have remarked that the Chase-Hershey work did not receive much notice at the time of the symposium. But the details contained within the writing by Hershey was enough to fully convince Watson, and later Crick, that DNA was indeed the genetic material.

The intriguing story, of how the mysterious secret of life, i.e., the DNA structure, was solved by Crick and Watson, was made wildly popular by Watson's 'tell all' book '*The Double Helix*.' In short, they found that the DNA structure had the sugar and the phosphate molecules on the outside of the double helical chain, like rails of a doubly twisted ladder, and that the nitrogenous bases, A, C, G, and T, made base-pairings A-T and G-C on the inside of the molecule.

In the years shortly after the discovery of the double helix structure of DNA was published, the predictions that arose as a result of the proposed Crick-Watson model and the structural nature of the model itself, were put to the test. Watson and Crick's DNA model structure has withstood the critical analyses of time.

3) Hershey and Chase worked with something called "bacteriophages." What exactly are these things, and how do they fit into the picture?

Bacteriophages are viruses that specifically target bacteria and infect mainly these types of microbes. The Greek term *phagein* means 'to eat,' and the name bacteriophage, therefore, means to 'eat bacteria.' Quite frequently, a bacterium that has been infected by a bacteriophage will be lysed, i.e., killed by being blown up.

Bacteriophages are highly specific to bacteria; in fact, certain bacteriophages may infect only certain strains of one species of bacteria. As such bacteriophages, or 'phages,' bind to bacteria, inject their phage-specific nucleic acid genomes into their bacterial hosts, and then direct the bacterium to cease making bacterial stuff and focus mainly on making new phage materials. Once all of the phage parts are synthesized, they may be processed a little using protease enzymes and then assembled into many copies of mature infection-ready virions. Lastly, the amplified viral particles are released from the bacterium and begin their cycle all over again by binding to new bacterial hosts.

Chase and Hershey utilized the *E. coli* bacteria-specific phage, called T2, a member of the T-even phages, which have viral heads and longer tails than the so-called T-odd phages. The viral head structure of phage T2 contains the double-stranded DNA genome. The T2 tails have binding sites for receptors located on the outer cell wall of the *E. coli*.

In essence, the phages have merely only two components: DNA and protein—the two primary candidates that were thought to be responsible for genetic inheritance. The big question was whether DNA or protein constituted the hereditary material.

The transformation work of the Frederick Griffith and the follow-up study by the scientific trio Oswald Avery, Colin MacLeod and Maclyn McCarty pointed to DNA as the so-called transforming principle, that is, that DNA was the heredity material. However, the proponents of the idea that protein was the heredity material effectively criticized the transformation work as not being able to say for sure that all or any contaminating protein (no matter how minute the amounts) was still lingering in each of their experiments. Avery, MacLeod and McCarty couldn't seem to shake off this one remaining criticism. Thus, the idea that 'genes were protein' remained.

This is where Chase and Hershey come into play. In simple terms, their T2 phage viruses contained only two components: DNA and protein.

Chase and Hershey reasoned that whatever part of their T2 phages (DNA or protein) entered the bacterium into the cytoplasm must be the heredity material.

4) The Hershey-Chase Experiments are often referred to as the "blender" experiments. Can you explain this?

Chase and Hershey used what is essentially the equivalent of a kitchen version of the Waring-made blender to detach radioactive T2 phages from the surfaces of *E. coli* cell walls. Investigators in the field had known at the time that phage T2 attaches to the outside walls of the *E. coli* cells. If Chase and Hershey could disrupt the attached T2 phage material that remained on the outside of the bacteria using the Waring blender, then the investigators could determine what material entered the bacteria—DNA or protein.

Their famous experiment is described briefly here.

First, Chase and Hershey made T2 phage radioactive. This was done in order to detect which part of the virus entered the *E. coli* bacteria during phage infection. In one flask, Chase and Hershey cultured *E. coli* in the presence of radioactive phosphorous (^{32}P), to radiolabel its DNA—you'll recall that DNA has phosphates as part of their double-helical chains. In another flask, they cultured *E. coli* in the presence of radioactive sulfur (^{35}S), to radiolabel its protein—you'll also recall that proteins are made up of amino acids, and sulfur can be found in the side chains of certain ones, like methionine and cysteine.

Next, Chase and Hershey infected the two batches of radioactive bacteria, each with phage T2—one *E. coli* batch that had ^{32}P-DNA and the other *E.coli* batch that had ^{35}S-protein. During the phage T2 infections of the radioactive bacteria, the phages eagerly took up the radioactive substances.

That is, the two infections produced phage T2 that were radioactively labeled. In one batch, they had T2 phage with ^{32}P-DNA, and in the other batch, they had T2 phage with ^{35}S-protein.

Then, Chase and Hershey infected new batches of *E. coli* with each of their radioactive T2 phages, one phage with ^{32}P-DNA and the other phage with ^{35}S-protein. Then, they used the now famous Waring blenders to agitate the infections by shearing off the emptied shells of radioactive viruses at various time-points during the infection process. Next, they centrifuged the mixtures of the two radioactive T2 phages and bacteria. The bacteria were pelleted to the bottoms of the centrifuge test tubes harboring on their cellular insides the injected (heredity) viral material, while the supernatants harbored that viral material left behind in the extra-cellular milieu.

Lastly, using a Geiger radiation counter machine, Chase and Hershey measured the amounts of radiation emitted in the supernatants, which represented viral material left behind and remaining on the outside of the bacteria, after being knocked off the cell wall surfaces of the *E. coli* by the blender.

Chase and Hershey observed that there was more ^{35}S-protein in the supernatant than there was ^{32}P-DNA. That is, the supernatant had almost 80% of the added ^{35}S-T2 protein sheared off the *E. coli* and into the supernatant after about 3 minutes of blending. In contrast, only about 30% of the total added ^{32}P-T2 DNA was removed from *E. coli* into the supernatant by 3 minutes of blending—the rest of the DNA was apparently inside the bacteria—unable to be sheared off by the blender because it was locked inside the *E. coli*.

Thus, it appeared that the vast majority of phage T2 DNA entered the bacteria while the T2 protein remained outside. These results were interpreted to mean that T2 phage protein encased the phage DNA and injected DNA into cells leaving the empty protein cases, called ghosts, on the outside of the bacterial cells.

Their data was published in the *Journal of General Physiology* in September of 1952. In an unprecedented practice, Martha Chase was a bona fide author of the famous work, and her name would forever go down in the history of science as being one of the key scientists of the legendary Hershey-Chase experiment. Textbooks dealing with genetics, biochemistry, and microbiology routinely describe the experiment in detail.

They concluded that the injected phage material was DNA. Hence, the heredity material had to be DNA!

While the injected DNA material was not 100% complete, as some radioactive protein was also found inside the bacteria, it was not enough to convince the scientific community at large that DNA was the genetic substance of life. To the later chagrin of the protein biochemists, it was enough, however, to convince Watson and Crick that DNA was the key to the secret of life. The believers that genes were made up of DNA were right.

5) Why are bacteriophages seen to be "model organisms" for research purposes?

The phages are indeed model organisms for conducting scientific research for several important reasons.

First, the phages were relatively straightforward to manipulate in the laboratory. They could be kept in refrigerators almost indefinitely while suspended in liquid broth.

Second, the phage bacterial hosts were almost just as simple to handle in the laboratory setting. One merely had to make broth in test tubes and culture flasks or prepare Petri plates with agar in them, in order to grow bacteria.

Third, the phages were easy to enumerate. One had only to mix the pure phages with bacteria in melted agar and pour onto an agar plate. The bacteria would grow to form a layer called a bacterial lawn. The phage would lyse the bacteria in the lawn, forming a hole in the lawn layer, called a plaque. Investigators could count the number of plaques formed in the lawn on the Petri plates and determine the viral titer in terms of plaque-forming units per volume.

Fourth, for both bacteria and their associated phages, mutational variants were easy to find and isolate. Such mutants, whether bacterial or viral, could be studied in detail using conjugation times and the blender to map the locations of the genetic determinants.

Importantly, bacteriophages were used early on in molecular biology as gene cloning vehicles. One of the earliest of these DNA cloning phages was called M13 virus, which readily infects certain strains of *E. coli*. Gene cloning into phage M13 facilitated the early efforts in DNA sequencing, which at the time required that the DNA template being sequenced to have a single-stranded version of it. Phage M13 readily fit this bill. Such efforts with the phages led ultimately led to improvements in sequencing efficiency and in the final elucidation of entire base-sequences of genomes for a great many organisms, including those of humans.

6) Hershey apparently shared the Nobel Prize for Physiology or Medicine with two other scientists- Max Delbruck and Salvador Luria - What happened to Martha Chase? Did she not receive any recognition?

Unfortunately, Dr. Martha Chase did not receive anything near the accolades that were enjoyed by Drs. Hershey, Delbruck and Luria. Chase did not garner a Nobel. In fact, historians of the sciences describe Chase as a tragic figure.

Martha Cowles Chase was born in Cleveland, OH, on the 30th day of the month of November, in 1927. At the time of the famous Chase-Hershey experiments, in 1952, Chase had been a recent college graduate and was Hershey's laboratory assistant at Cold Spring

Harbor, in NY. She had earned her BS degree from the College of Wooster in 1950. While Hershey was the lab chief, Chase was the person who actually performed the now famous Chase-Hershey T2-*E.coli* experiments.

Because of her technical and scientific acumen in the laboratory setting, Chase was invited to work at Oak Ridge National Laboratory in TN in 1953, after she completed her T2 phage work in *E. coli*. Next, while working briefly at the University of Rochester, NY, Chase would spend each summer participating in the very famous Phage Group discussions. In 1959, Chase enrolled in the Ph.D. program at the University of Southern California. She took her doctorate in genetics in 1964.

Regrettably, Dr. Chase developed early onset dementia in the late-1960s, losing some of her short-term memory and eventually causing her to lose her job and end her scientific career. Further, to make matters worse, Chase had to stand by and watch Hershey get the Nobel and was never mentioned by name even once in Hershey's Nobel lecture, delivered on the evening of December 12, 1969. Chase died on the 8th day in the month of August in 2003, at the age of 75.

7) Martha Chase was also a geneticist. Did she make any contributions to that area?

Dr. Chase's primary and widely known contribution to genetics was in personally conducting the famous Hershey-Chase blender experiments, which showed that phage T2 DNA was the biological material that entered into the host *E. coli* bacteria, thus, strongly implying that DNA was the genetic material.

Dr. Chase conducted further experiments while at Rochester in which she and colleague A.H. Doermann examined genetic recombination in phage T4 and genome replication in the bacteriophage. They studied how closely linked genetic elements along the T4 chromosome did not seem to undergo genetic crossover as frequently as those genes that were located further apart along the phage genome. They also noticed that genetic recombination accompanied DNA replication during T4 phage progeny production.

This work was important because other investigators were able to extend the recombination-replication connection to *E. coli* and later to many other organisms. In fact, human DNA replication involves recombination events during DNA replication. I think their genetic work was key also to understanding physical gene mapping, another important advance in the field of genetics.

8) What have I neglected to ask?

Many of Hershey's scientific colleagues who knew him personally, have often written in their personal memoirs about the so-called notion of 'Hershey's Heaven.' While individual accounts vary in their details, the main gist of the phenomenon refers to the intense satisfaction inherent in performing experiments in the laboratory or in making key important discoveries. Hershey was said to have mentioned this to visiting dignitaries and colleagues. I have to admit myself that some of my own happiest scientific moments were experienced at my laboratory bench conducting experiments and in writing about my own scientific discoveries.

There is a sense of not only cherishing deep satisfaction in making new scientific discoveries that perhaps no one else in the world has done, but also in knowing that there's a good chance that your discovery may somehow be important in making the world a better place. In a most agreeable way, many scientific endeavors have helped to make a prettier world for all of us.

9) What is the relevance today of the Hershey-Chase experiments?

Many writers of science have wondered why the not-so-clear-cut data generated by the famous Hershey-Chase blender experiment clearly convinced a certain few key investigators, but not many others, of the potential for DNA to hold the hereditary role in biology. The fact remains, however, that shortly after the publication of the Hershey-Chase paper in 1952, the DNA structure had been largely solved by Watson and Crick. The rest of the world took it from there. The scientific works of Frederick Griffith and of the trio Avery, MacLeod and McCarty, were at the time (1940s) not enough to convince the 'protein geneticists' while the Hershey-Chase data were indeed effectively sufficient to do so in the 1950s.

The elegance of the Hershey-Chase experiment is literally one for the history books, as well as for the science books. Their work stimulated an unprecedented body of work all dealing with DNA, the secret to life. The Hershey-Chase experiment paved the way to the DNA structure, DNA recombination technology, gene cloning, DNA genetic manipulation, DNA and protein sequencing, genome determination projects, protein structure, protein biochemistry, and bioinformatics, to name only a few.

Chapter 29

JONAS SALK

1) I daresay that most schools and most science curriculums mention the name of Jonas Salk somewhere along the line. First, when and where was he born and where did he get his training?

I believe you have quite accurately described the widespread and current interest in teaching this topic in all levels of schools. Hence, you have indicated the universal importance of Dr. Jonas Salk, discoverer of the polio vaccine, which played a critical role in altering the course of infectious disease history. Jonas Edward Salk was born on the 28th of the month of October, in the year 1914 in East Harlem in New York City, NY, to Jewish parents, Daniel Salk, his father and a struggling businessman, and his mother, Dora Press, a Russian immigrant, who handled the family's finances. Jonas was the eldest of three siblings. The young child Jonas had witnessed the devastating effects of the Great Influenza pandemic, just prior to 1920, and of the many children his age with leg braces, the latter presumably from the effects of the potentially debilitating disease poliomyelitis.

The child Salk attended public school number 44, located in the Bronx, New York. His early brilliance was already noticed by his teachers, who arranged for Salk to skip into higher grades. He was a deeply curious child, an avid reader, and greatly enjoyed learning. In 1926, when young Salk was 12 years of age, he was moved to a special and competitive public high school, called Townsend Harris, for talented and gifted children. Salk was reported to have been a serious student while at Townsend Harris, which had a pre-college preparatory curriculum for students.

Earning a competitive college scholarship at the age of 15, Salk was admitted to City College of New York (CCNY) and started attendance in 1929, planning to enter law school and, thus, enrolling in pre-law courses. Salk was the first member of his immediate family to have gone to college. Surprisingly, Salk's first semester did not end well, earning poor grades in several courses. His mother was reported to have said that Salk would not have made a good attorney, anyhow. Although he had earned a relatively poor grade in his freshman Chemistry course, he became interested in the topic.

During his sophomore year of college Salk switched his major to chemistry with a pre-medicine track, and his grades improved but only slightly. Inspired by the reading of a biography of the great Louis Pasteur, undoubtedly one of the greatest scientists of all time, Salk made the decision to conduct medical research for a career, rather than practice medicine. Salk graduated from CCNY with his Bachelor of Science degree in the area of Chemistry, in 1934.

Planning to combine basic science with medicine, Salk was admitted to Bellevue Hospital Medical College, which was later named the New York University College of Medicine, and started medical school in 1934. This is the same school that the famous Walter Reed had attended during the late 1860s. During Salk's tenure the curriculum consisted of 2 years of medical science courses, such as bacteriology, human anatomy, pathology, pharmacology, and physiology, followed by 2 years of clinical experience in the hospital. Excelling in many of his courses, Salk's acumen was noticed by one of his teachers, Dr. Keith Canaan, a prominent biochemistry professor, who offered Salk a fellowship to perform research in Canaan's laboratory. Thus, Salk took a year off from medical school to conduct research. He studied egg albumin protein denaturation, which was a protein biochemistry project; next, Salk studied antibody production in rabbits in response to the bacterium *Streptococcus pyogenes*, the causative agent of scarlet fever and other diseases. In this latter project, Salk discovered a new way to harvest the bacteria from large broth volumes, which turned out to be a new development worthy of publication. Thus, Salk published his very first paper for this work in a scientific journal, at the age of 22.

In 1936, Salk returned to medical school to continue his studies. Presumably while in his immunology course, Salk became perplexed with two seemingly contradictory concepts. On the one hand, he was taught that effective immunization to the bacterial diseases tetanus and diphtheria could be accomplished by first denaturing their respective bacterial toxins beforehand and then injecting them into individuals. On the other hand, he was also taught that in order for immunization against viruses to work properly, one had to actually be infected with a live virus and that dead viruses did not properly achieve immunization. At the time, this notion was certainly true with the then known vaccines directed against small pox, developed by Edward Jenner, rabies, developed by Louis Pasteur, and yellow fever, developed by Max Theiler.

Salk's curiosity about bacteria, viruses, and vaccines led directly to his entry into another research laboratory, this time that of the newly appointed chair and professor of bacteriology, Dr. Thomas Francis, Jr., a virologist and discoverer of a new influenza virus strain. Back in the laboratory of Dr. Francis during his last year of medical school, Salk provided experimental evidence supporting the hypothesis that killed influenza virus provided immunity in mice by measuring antibody production. This was a turning point for Salk.

He finished his medical studies, earning his M.D. degree in 1939. Immediately after graduation the newly minted Dr. Salk married Donna Lindsay, whom he had met in Woods Hole, MA, where he had been conducting research during the summers between medical school years.

Salk became an intern at the Mount Sinai Hospital providing clinical patient care and conducting surgery. It is during this time that Salk, influenced by Donna, partook in political activities that would come back to haunt him in later years, as his FBI file has documented during their investigations of him and his wife. In 1942, with Salk's internship at an end and having trouble securing a permanent position, some of the trouble being ingrained anti-Semitism, Salk had received a post-graduate fellowship from the National Research Council to work with his former mentor, Dr. Francis, who had since moved to the School of Public Health housed at the University of Michigan, studying influenza immunity, as a research associate. Salk travelled to post-war Germany to investigate the efficacy of an influenza vaccine in the American troops stationed there.

In 1946, Salk was promoted to the rank of assistant professor at the University of Michigan, and in 1947, Salk moved to the University of Pittsburgh as an associate professor and director of their Virus Research Laboratory. It was also Salk's first position in which he had his own research laboratory dedicated to the study of virology.

In 1949, Salk became full professor.

2) Polio used to strike fear into the hearts of many, but as I understand, it is almost non-existent (except for a few cases here and there). When was polio most common, what were the symptoms, and how did we get it under control?

The disease commonly called polio has the technical name of poliomyelitis among clinicians, and historically it has been called infantile paralysis. In the U.S. cases of polio reached a maximum in the early to mid-1950s (peaking at 37 cases per 100,000 in 1952 in the U.S.), and on a world-wide scale polio was in a full pandemic status, peaking sometime during the early 1960s. During these years, children were banned by their parents from playing outdoors during summers because of the pervasive fear of contracting the potentially debilitating polio.

The mode of polio transmission includes the so-called oral-fecal route, brought about by poor hygienic practices, including the consumption of food or water that are contaminated with the virus. The symptomology of polio has varied in severity, depending on the extent and corresponding severity of the viral infection. In people who are not vaccinated, there are four possible outcomes for those who become infected with the poliovirus microbe.

The first outcome is the mildest and is manifested by having no symptoms. This asymptomatic outcome is the most common, occurring in the vast majority (about 90%) of those individuals who are infected. In these cases, the poliovirus is stopped in its tracks in the throat or the gut by components of the person's immune system.

The second outcome of poliovirus infection is called abortive poliomyelitis. This outcome occurs in about 5% of infected patients and manifests itself as a minor illness, involving symptoms like fever, fatigue, headache, sore throat and vomiting. These symptoms may appear three to four days after making contact with the poliovirus.

The third outcome is called non-paralytic poliomyelitis and sometimes aseptic meningitis, and it happens in about 1 or 2% of patients who are infected with the virus. In these cases, the poliovirus will have progressed onto the central nervous system (CNS) and the membrane lining that covers the CNS, called the meninges, of the patient. The symptoms include those found in abortive minor illness, plus new symptoms that include muscle pain, often in the back, and muscle spasms.

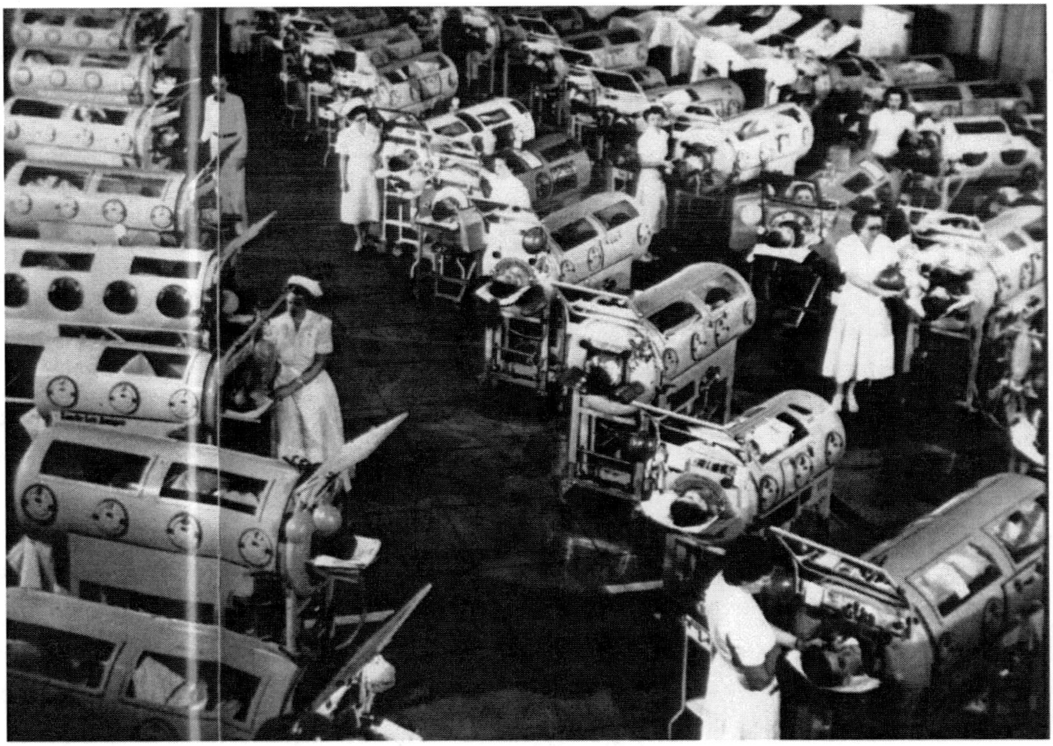

Figure 1. Iron lungs used for polio patients.

The fourth and last outcome of poliovirus infection is called the paralytic polio, paralytic poliomyelitis, or major illness. This particular outcome is the most severe and the least common, occurring in roughly 0.1 to 0.2% of infected patients. The patients will have exhibited the symptomology of the minor illness first, and after it has come and gone, usually about 3 or 4 days, paralysis occurs. The extent of the paralysis is widely variable, ranging from mild and temporary to severe and permanent, depending on the extent of the viral infection in the patient. In these severe cases the virus will have entered the neurons of the spinal cord and brain, involving flaccid paralysis of one or more of the extremities; if recovery occurs it may take anywhere between 6 months to 2 years.

In the severest case of paralytic polio, called bulbar poliomyelitis, additional muscles are affected, including muscles of the throat, voice cords, and lungs; this condition may be fatal if the so-called iron lungs or modern respirators are not used to help patients breathe. The images of children residing in iron lungs during the early part of the 20th century are compelling.

In older patients who have survived a polio infection, occasionally they exhibit a condition known as the post-polio syndrome; here the neurons that were destroyed by the virus now have an effect on the muscles which had been (but are no longer) innervated by the destroyed neurons, causing corresponding muscle deterioration. In these cases, the virus is not in the patient as the immune system has eliminated it long beforehand.

Control of polio has been brought about primarily in two ways. The first is improved hygienic conditions which prevent the fecal-oral transmission mode in developed countries. One problem with this effort has been access to clean drinkable uncontaminated water and food in developing countries. The second way is prevention by use of vaccines and a vast vaccination program. Bringing the control of polio about on a worldwide scale has been one of the most successful triumphs in the history of modern biomedical science.

3) I surmise that there is some type of vaccine that prevents polio- or am I wrong?

You are absolutely correct. There are two principal vaccines that have been in use. The first vaccine was developed by Dr. Salk in which he inactivated the polioviruses by chemical treatment with formalin and is called the inactivated poliovirus vaccine (IPV).

The second vaccine was developed by Dr. Albert Sabin in which he attenuated the polioviruses into a non-pathogenic form by multiple passaging of the virus in cell culture from generation to generation. The attenuated poliovirus is live, and the vaccine is taken by mouth. Thus, Sabin's vaccine is called the oral poliovirus vaccine (OPV).

Both vaccines are directed against all three types of poliovirus strains or serotypes, called Types 1, 2 and 3.

Because of its ease of administration and its lower costs, the OPV had supplanted IPV in the global eradication efforts.

4) Now, to clarify, what is an IPV vaccine, and why is it important?

When Salk developed the IPV he did it by first growing the three strains of poliovirus, called Types 1, 2, and 3, in monkey kidney cells, and later in the more famous HeLa cells, in culture. Next, the intact whole polioviruses were purified from the cultured cells and then soaked in the same chemical that's used to preserve corpses, a mixture of water and formaldehyde, called formalin. The formalin-treated polioviruses are then used for inoculations. The IPV inoculations are administered by 4 intra-muscular injections with a needle at 2, 4, and 12-18 months of age and then lastly between 4 to 6 years of age.

The IPV is important because it was the first effective polio vaccine. Furthermore, the IPV was shown to stem the tide in the polio epidemics and was not involved in causing vaccine-related polio cases. IPV is also currently the routine polio vaccine of choice in regions where the wild-type polioviruses are absent.

5) As I recall from history classes back in grade school, Franklin Delano Roosevelt was often seated and was rarely seen standing. I believe he had polio- and apparently, it impacted him his entire life. What do we know about the course of polio- when does it first appear, and what are the signs and symptoms?

If I may be so bold as to say so, I am convinced, as probably many historians of science are, too, that had Franklin D. Roosevelt not had a case of the paralytic polio, the success we've enjoyed in vastly reducing its incidence and prevalence might never have happened in modern times. FDR was a key figure in the efforts that were generated for the purpose of someday eradicating polio from the face of the Earth. FDR put the issue to the forefront of society's funding and scientific machine to address the polio disease epidemics and ensuing pandemic.

The process for FDR's involvement began during the month of August in the year 1921, when he started to exhibit the signs and symptoms of polio while on vacation in his family's summer home at Campobello Island, located in New Brunswick, Canada. Prior to the onset of polio, FDR had been a senator, assistant secretary of the navy, and had just lost the race for vice-president. The outcome for Roosevelt was that he was paralyzed from the waist down in both of his legs and could not walk. Of all of the difficulties that had been plaguing FDR at the time, contracting paralytic polio was an extremely problematic one, as it was potentially a career-ending situation. Being in the public eye it was unheard of for paralyzed individuals to be running for high office, and in particular Roosevelt had his eye on the U.S. presidency.

Perhaps wishing to provide encouragement, his doctors had informed Roosevelt that there was a good chance of recovery and that he might walk again. Consequently, FDR was forever optimistic that recovery was at hand, and he made every effort to improve his chances for recovery, with physical therapy and any other kind of therapy, no matter how futile the putative therapy might be for him. He never gave up hope.

Meanwhile, until Roosevelt could make a full recovery, if one was to be had at all, that is, he also made every effort to hide his paralysis from the public. He banned the taking of photographs with him in a wheelchair, the telltale sign of being an invalid. Any photographs taken were of him always sitting behind a desk, his wheelchair out of view, or of him sitting in his car, saying a few words to the audience and the press. If he were to give a speech, he had colleagues accompany him to the podium as he wobbled his way. Then, once secure at the podium, he could prop himself up with his arms and deliver his speech. He had a good system going for him.

FDR made it possible for funding, research and healthcare for polio to be pushed to the front burner. He established centers for treatment, physical therapy, and research. The March of Dimes program encouraged all children to mail a dime in an envelope to the foundation—this was a highly successful program. This is also where the push for vaccines and sanitation, etc., were all put into motion. Within a few years the Salk vaccine, and later the Sabin vaccine, became available, although too late for president FDR, who died while in office on the 12th of April in 1945, just before the end of World War II.

The signs and symptoms of polio often appear anywhere between 3 and 6 days after exposure to the poliovirus. Sometimes the minor illness symptoms will disappear, and 3 or 4 days later the classic poliomyelitis symptoms set in. The virus often enters the patient after consuming food or water that's contaminated with the polio-causing virus. The virus may infect the cells of the throat, causing a sore throat. The virus may end up in the gastrointestinal tract, causing a GI tract disturbance. The throat and the gut are most often where the virus is stopped by the patient's immune system. If the virus progresses past these stages to the nervous system, like the CNS, more severe cases can occur. Paralysis may be observed as soon as a week after exposure to the virus. The virus will be shed from the patient in the stool, and it may be a source of transmission to the next patient.

Interestingly, polio disease made one of its first appearances in recorded history as an ancient Egyptian hieroglyph. The hieroglyph of the 18th dynasty, dating between 1580 – 1350 B.C. shows an unidentified man with a withered right leg indicating a "drop foot" syndrome, pointing to an asymmetric paralysis typical of polio. The leg has atrophied, most likely due to loss of innervation as the motor neurons may have been damaged or destroyed during poliovirus infection.

Figure 2. Ancient Egyptian hieroglyph suggesting the occurrence of poliomyelitis.

Speaking of appearances, I think it is encouraging also to speak of polio's disappearances. In 1979, all three strains of poliovirus disappeared from the U.S., all due to the vaccinations made possible by Salk and Sabin. In 1999, Type 2 poliovirus officially disappeared from the Earth—we've not seen polio cases attributable to this strain ever since. Likewise, poliovirus Type 3 disappeared from the world in 2012. The last poliovirus strain Type 1, the worst one of the three, remains in small pockets, with a few hundred cases per year being recorded during the period between years 2001 and 2015. When that day arrives in which all polio disease will be totally gone from the world remains unknown as of this writing. So far, only smallpox has been eradicated from the planet due to human intervention. Polio is the second contagion that's slated for a similar type of world eradication.

6) What have I neglected to ask and is there a contemporary treatment for those, for whatever reason, are afflicted by this disease?

While there are currently no useful anti-viral medicines that target the poliovirus, *per se*, medical approaches to polio have primarily focused on prevention, with the use of the Salk and Sabin vaccines and improvements in interrupting the fecal-oral route of polio transmission. There has been little need for the iron lungs for many years because of the massive vaccination programs that were established in the 1950s. If an iron lung were needed modern day respirators which are much less cumbersome would suffice in cases of bulbar polio. According to the Guinness World Records, Ms. June Margaret Middleton holds the world record for time spent living in an iron lung—at least 60 years.

7) What have I neglected to ask?

Sixteen year old Bill Kirkpatrick was the first person ever to have received the Salk vaccine. Salk himself had injected the eager volunteer in June, of 1952. In fact, Salk had injected the first 43 volunteers and then had terrible trouble sleeping afterwards, agonizing over what might happen.

Although Salk demonstrated great acumen in making the vaccine happen on a scientific basis, he was reported to have been severely out of his element during the grand announcement of the success of the vaccine trials, by making a huge blunder. In short, he thanked everyone he could think of, except those individuals in his lab who actually made the vaccine and conducted the vaccine trials. What's worse is that he repeated his blunder time and time again, never seeming to quite convey the credit to his immediate colleagues.

I briefly mentioned the use of HeLa cell cultures for the polio vaccine development; these cells are now of current and wide interest, after the publishing of a book by Rebecca Skloot about the origin of these very important cells. Sadly, the cells were taken from Henrietta Lacks without her permission or knowledge after she died of cervical cancer. The HeLa designation comes from the first two letters of her first and last names. The cells were of immense importance for biomedicine because of their capacity for indefinite passaging between generations of cell growth. This greatly facilitated the development not only of polio vaccine but for a great deal of advances in both basic and applied research. Today, if biotechnologists or any other investigators wish to use HeLa cells, they need permission from her family. The first publication with the Lack family having provided permission was the study in which the HeLa genome DNA sequence was determined. A film about the "*Immortal Life of Henrietta Lacks*" was released in 2017.

Although Salk was widely celebrated in his later years with numerous tributes, awards, and accolades—he even had an entire institute named after him, the Salk Institute in La Jolla, CA, he never received the Nobel Prize, and neither did Sabin, for that matter. Salk died on the 23rd of June in the year 1995.

Chapter 30

FRANCIS CRICK, JAMES WATSON AND ROSALIND FRANKLIN

1) Dr. Varela, we hear it all the time- those three initials- DNA, and I learned it back in high school biology. But just for the record, what exactly is DNA (and what differentiates it from RNA?)

The term DNA is actually an abbreviation for deoxyribonucleic acid, but the initialed form, D.N.A., has been largely ignored over the many years since its structure was determined. This central life molecule has been referred to in many ways, such as the molecule that holds the secret to life, the blueprint for life, the hereditary material, the genetic code, genetic information, etc. The maxims are aplenty. The importance, however, of this nucleic acid molecule to all living beings on Earth cannot be overstated.

From a chemical or biochemical perspective, the structure of DNA has three general pieces to it. The first part is actually a sugar, commonly known as deoxyribose, which is composed of the atoms carbon, hydrogen and oxygen. In the sugar there are 5 carbon atoms, and each atom is numbered 1' through 5', pronounced "one prime" and "five prime," respectively. The "deoxy" part of the deoxyribose sugar means that the molecule does not have an oxygen atom attached to it, at carbon number 2' of the sugar molecule. That is, the oxygen is missing from the second carbon of the sugar, hence, the name deoxyribose. The sugar is also in a so-called ring form, the technical name being a β-furanose form, in which the molecule circles back upon itself using 4 of the carbons and one of the oxygens to form the ring; so in general the deoxyribose sugar falls into a class of carbohydrates called pentoses, because of the 5 atoms that make up the sugar.

The second part of DNA consists of a series of 4 chemical base compounds, each with nitrogen atoms embedded in them, often called nitrogenous bases, or just simply bases. Two of the four bases, called guanine (G) and adenine (A), originate from a class of molecules called purines. On the other hand, the other two of the four bases, called cytosine

(C) and thymine (T) derive from another class of molecules called pyrimidines. The names have interesting origins. For instance, guanine was extracted from bird manure called guano whereas thymine was extracted from thymus tissue. If the four bases are each attached to sugar, usually at carbon number 1', then these complexes are called nucleosides, and each of these four sugar-base nucleoside complexes are now specifically called deoxyguanosine, deoxyadenosine, deoxycytidine, and deoxythymidine (or thymidine).

The third and the last part of DNA considered here is called a phosphate, which has a central phosphorous atom linked to four oxygen atoms. The phosphate molecule is attached to carbon number 5' of the sugar using one of the oxygen atoms to make the connection. Hence, when all three of these structural parts are attached to each other, they are then called nucleotides, and these four base-sugar-phosphate nucleotide complexes are now each specifically called deoxyguanylate, deoxyadenylate, deoxycytidylate (or cytidylate), and deoxythymidylate (or thymidylate). The nucleotides will make up a sequence of bases along the DNA chain.

The arrangement of these three parts within a DNA molecule are organized with the sugar-phosphates on the outside forming two long chain backbones, like ladder side rails, and the bases on the inside, like ladder rungs or steps. Each of the sugar-phosphate backbone rails are known as strands, and since there are two of these sugar-phosphate strands, the DNA molecules are called double stranded. Inside the double helix, the bases bind to each other using weak hydrogen bonds in which A binds to T, called an A-T base pair, and in which G binds C, forming a G-C base pair. If one knows the base sequence of one strand then you'll know the sequence of the other strand—this phenomenon is known as complementary base-pairing; it's also known as Watson-Crick base-pairing.

One strand will go in one direction, called a 5' to 3' direction; while the other chain, attached to the first chain by forming hydrogen bonds between the bases, will go in the opposite direction, called the 3' to 5' direction. This phenomenon is known has anti-parallel. This anti-parallel nature of the DNA structure was a critical finding when the DNA model was proposed by Crick and Watson in 1953. The sugar-phosphate backbones on the outsides of the molecule and the bases on the inside in their ladder formation, are twisted about to form the famous double helix structure of the DNA molecule.

Structurally speaking, the RNA molecules differ from their DNA counterparts in several general ways. First, the sugar in the RNA molecule is composed of ribose (instead of deoxyribose like in DNA) with its oxygen atom present on its carbon number 2'. Second, the bases differ in that RNA has uracil (U), instead of thymine, both of which are pyrimidines. The ribonucleoside, composed of a ribose-uracil complex, is called uridine, and with the phosphate attached to the 5' carbon of ribose, it is then called uridylate. Often, RNA molecules are single-stranded, but can fold about themselves to make functional three-dimensional shapes, occasionally even forming double stranded versions of the RNA molecules.

Functionally speaking, the nucleic acids DNA and RNA differ from each other in several other ways. Let's talk about DNA first. The string of the four bases (G, A, C, T) along the double helical DNA molecule is often called the nucleotide base sequence or simply the DNA sequence. The hereditary material stored in DNA is harbored in its base sequences and contained within certain genetic elements known as genes. The genes on DNA encode the blueprints for proteins that are needed for life to occur. DNA copies itself by making new DNA molecules with largely the same base sequences, in a process called DNA synthesis or DNA replication. The replicated DNA, for example, will then go into a subsequent generation of living organisms. DNA has essentially two so-called template functions: one for making DNA copies (replication) and the other for making RNA transcripts (transcription).

Let's consider RNA for a moment. The genetic information that's stored in DNA is then transferred to RNA, in a process called RNA synthesis or transcription. Essentially, three kinds of RNA molecules have been discovered over the years: messenger RNA (mRNA), ribosomal RNA (rRNA), and transfer RNA (tRNA). The mRNA molecule, also called a primary transcript, harbors the information needed to specify the production of protein, a process called protein synthesis, or translation. The code message from DNA that's now in the mRNA, called codons, is read by the cell's translational machinery like, for instance, the ribosome plus associated molecules. The tRNA "transfers" amino acids from the cell's cytoplasm to the translational apparatus; there are dedicated tRNA molecules for each of the 20 amino acids; the anti-codons on the tRNA molecules are complementary to the codons on mRNA transcript. The rRNA makes up a large part of the gigantic ribosome, along with many proteins. Interestingly, and perhaps shockingly, the 'enzyme' which binds the amino acids together is made up of RNA! It's the rRNA that's the enzymatic part of the translational machine. Most people usually think that enzymes are proteins, but in this case, the key enzyme is rRNA.

The translation machinery uses all three RNA types in order to produce proteins, which are chains of amino acids linked together with peptide bonds and are folded into three-dimensional structures to form enzymes and other proteins, which then carry out life's processes, such as metabolism or cell growth, etc.

The genetic information that's stored in DNA flows to DNA for the next generation and to RNA and to protein. In the 1950s this was known as the Central Dogma (because it was a central idea about life but with little or no experimental data to back it up, at the time), but in modern times it is firmly established and is now called genetic information flow.

2) Why is it important and what is its major relevance?

Simply put, without DNA there would be no life as we presently know it. DNA is necessary for life to occur. DNA is the center of all life, including bacteria, fungi, plants, trees, animals, and human beings. DNA specifies life. Every living cell has a copy of all of the DNA that specifies the living being. Even viruses, which investigators are not necessarily sure are alive, have a nucleic acid core; in some viruses, the viral nucleic acid is RNA, while other viruses have DNA.

DNA is important because it manages to copy (replicate) and then convey its hereditary genetic information that's stored in the DNA to the next generation of living beings (progeny) and to all subsequent generations (descendants). Along the way, from generation to generation, DNA changes its base sequences in processes called mutation and recombination and thus results in the evolution of all living beings. Evolution may thus be defined as the changes in DNA that occur from generation to generation. This is why progeny appear somewhat different from their parents and not identical.

DNA is important for many reasons. It's important medically because, for instance, in genetic diseases dysfunctional or non-functional proteins may be made instead of the normal functional proteins. These aberrant proteins may then cause other life biomolecules, cells, tissues, organs, or physiological and biochemical processes to malfunction, possibly causing suffering, disease and a loss in the quality of life. It may shorten the lifespan of the individual person who has a genetic disease.

Another major reason why DNA is relevant is because investigators have invented various technologies to manipulate it for good. One such method is called recombinant DNA technology. The method has been used, for instance, to clone genes for both basic and applied studies. Gene cloning allows study of life's process and production of useful products. Another method has been to determine the base sequences along the DNA double helix.

For example, medical tests can be developed to diagnose disease, whether it's genetic or infectious, etc. DNA can also be manipulated by genetic engineering to produce useful products, like medicines, food crops that resist insects or ripen later to assure freshness, vitamins for making our biochemical reactions work efficiently, immune system development to fight cancer or infections, insulin for diabetes, vaccines for prevention of disease, or to make the normal versions of proteins that are aberrant in genetic diseases for replacement therapy. There's even been an advance in the technological DNA engineering development, called gene therapy, which is designed to use non-pathogenic viruses to insert the correctly functioning DNA gene into a person with a genetic disease. A newer technology has emerged recently, called CRISPR (pronounced "crisper") that's designed as a gene editing system; it can insert a gap mutation within a targeted gene to inactivate it in order to learn what the gene's function encodes, or the DNA gap can be substituted with another gene, perhaps the normal gene. Now, if you must know, CRISPR is another acronym, and it stands for clustered, regularly interspaced short palindromic repeats.

The study of DNA base sequences also has led to the development of numerous genome projects in which the entire genetic component, the genome, the collection of an organism's entire gene make-up, can be determined. The new field, called genomics, allows study of gene function, genetic and species composition of an environment, and allows comparisons of genomes to follow evolutionary processes. Genomics has led to the development of proteomics, or the entire collection of proteins in a given system, like an organism or group of related organisms. DNA sequence technology has been enhanced to such a great deal that our computer hardware and software cannot keep up with all of the DNA data stored in computer databases. Thus, the new field, called bioinformatics, has emerged in which knowledge of computers is combined with that of biology, in order to be able to design new computers that are able to "crunch" the new ever-accumulating DNA genome sequence data.

DNA gene sequences can thus be compared, transcribed into RNA, translated into proteins, and protein sequences can be deduced straight from DNA so that protein structures and functions can be studied. These new fields are important in the biomedical sciences for the benefit of human health, such as immunity or study of new cancer treatments.

3) James Watson, Francis Crick, and Rosalind Franklin- where were they working and how did their discovery of DNA structure come about?

Interestingly, Drs. Franklin, Crick and Watson did not discover DNA. On the contrary, the discovery of DNA actually happened much earlier in 1869 when Friedrich "Fritz" Miescher had isolated it from the pus of infected bandages as a phosphorous-containing substance he called 'nuclein.' The material had been derived from the nuclei of the white blood cells in the pus-laden bandages.

Instead, the work of Franklin, Crick, Watson and others such as Maurice Wilkins and Raymond Gosling, led to the elucidation of the STRUCTURE of DNA—that is, that the bases were on the inside with A binding to T and G binding to C, and that the sugar-phosphates being oriented on the outside, making strands that went in opposite directions and twisted about in a double helix configuration, etc. This was their history changing discovery—the DNA structure.

It was correctly believed at the time that knowledge of its structure would lend insight into its function.

At the time of their great discovery of the DNA structure elucidation in the early 1950s, Crick and Watson were housed in Cambridge, England at the prestigious Cavendish Laboratory, under the direction of the Nobel laureate Sir Lawrence Bragg. Franklin, Wilkins and Gosling were in London, England, at the famous Kings College, in their new biophysics department under the direction of Prof. John Randall.

Crick and Watson came about their discovery by using nucleic acid chemistry knowledge gleaned from the scientific literature and, interestingly, from the precise and carefully collected X-ray crystallographic diffraction data of DNA collected by Franklin and Gosling—the famous so-called "photo 51," which had been, up to that time, the best diffraction data ever seen of DNA. It's considered to be a classic and elegant image that's often included in textbooks today. The story is often told that this diffraction data in Franklin's report, was shared with Crick and Watson by Wilkins, and this led them to correctly propose their model of the DNA structure.

Together, Crick and Watson built a molecular model of the DNA structure showing the main atoms and all of the correct bond angles and arrangements of the sugars, bases, and phosphates, complete with its double helical nature of its configuration. All investigators published their papers separately in the same issue of the scientific journal *Nature* on April 25, 1953.

These seminal *Nature* papers ultimately led to a revolution in the history of DNA. It led to the development of the field of molecular biology, which is, in one sense, the study of how DNA replicates itself and transcribes to make RNA which is translated to make protein. In another sense, the discovery of the DNA structure immediately led these investigators to gain tremendous insights into function. For instance, its complementary nature led them to propose a mechanism for DNA replication. Lastly, the DNA structure elucidation led to the field of molecular biology as a methodological approach to the study of living beings at the molecular level in a way that allowed the elucidation of their physiological, biochemical, cell biological, immunological, neurological, pharmacological, microbiological processes, etc. In fact, virtually all fields within the biological and biomedical sciences are using molecular biological approaches for their studies.

4) What do we know about these three individuals?

Rosalind Elsie Franklin was born on the 25th of July in 1920, in Notting Hill, England, to Jewish parents Ellis Franklin, her father, and her mother, Muriel Waley Franklin. The Franklin family was a prosperous one, and the child Rosalind was knowledgeably fortunate in that her family was well-traveled, especially to foreign countries, a practice that expedited the learning process for an intensely curious and gifted child. Interestingly, the family has frequently told the story that, in 1910, a member of the Franklin family, Rosalind's uncle Hugh on her father's side, either tried to or indeed had accosted Winston Churchill, who was then England's Home Secretary and staunchly against the suffrage movement. Apparently, Churchill was unharmed in the incident.

Young Rosalind was enrolled in Norland Place where she was recognized early on as being extremely bright. Franklin later attended The Lindores School for Young Ladies till the month of December in 1931. In early 1932, Franklin then attended St. Paul's Girl's School, in west London, and she was known to be an outstanding student and was bestowed numerous academic awards. She studied French, German and Italian languages plus math and science. Unfortunately, however, biology was not taught to females at the school—only to males. Franklin attended college at Newnham, in Cambridge, starting in 1938. At the time, females at Cambridge University were not accepted as *bona fide* members of the university. Instead, they were considered students of Newnham College, Cambridge—the females-only component of Cambridge University.

Nevertheless, as a college student concentrating in the Natural Sciences Tripos (called Tripos because students taking honors exams sat on stools with three legs), Franklin read Pauling's famous book *The Nature of the Chemical Bond*. Franklin learned physics, virology of the tobacco mosaic virus, protein chemistry, and acquired a keen interest in crystallography, especially in the method of X-ray diffraction. Despite having a propensity for self-doubt and being a college student in England during the war years and the blitz, Franklin took her undergraduate degree in the field of physical chemistry in 1941 with very good grades.

Franklin's academic record qualified her for a graduate fellowship at Cambridge, where as a Ph.D. student, she entered the research laboratory of Ronald G.W. Norrish, professor of physical chemistry, who was later to garner a Nobel. Unfortunately, Norrish at the time was possibly an alcoholic and, thus, he had trouble acquiring a security clearance to do war-related research work. Furthermore, he assigned a trivial and undoable graduate project for Franklin—it was Franklin herself who discovered the fundamental error in the project's design.

Hence, a row ensued between the two, and in 1942 Franklin moved to Kingston, England, taking a government job as an Assistant Research Officer at the British Coal Utilisation Research Association (BCURA). In her new post, personnel were allowed to use their work towards the Ph.D. degree; the added benefit was Franklin's work helped the war effort. At BCURA, Franklin studied how bituminous and anthracite coals were permeable to various substances, like helium gas or water.

Successfully completing her Ph.D. thesis in mid-1945, the topic dealt with the physical chemistry of solid organic colloids in coal and related compounds. Franklin's first publication in a scientific journal in 1946 was based on her Ph.D. work in which she provided experimental data to support her hypothesis about a molecular sieve property attributable to coal materials.

After acquiring her doctorate degree in physical chemistry, Dr. Franklin furthered her career in February of 1947 by accepting a purely research position as postdoctoral fellow in the laboratory of Jacques Mering at the Laboratoire Central des Services Chimiques de L'Etat, a laboratory that was housed in post-war Paris, France. It was a scientifically productive time for Dr. Franklin.

In the Mering lab, Franklin honed her skills in the technique of X-ray diffraction in studying clay and coal materials. The methodology was cumbersome with high maintenance equipment, sophisticated mathematics and exposure to potentially toxic chemicals. Franklin nevertheless published several key papers in the field, including a paper in the prestigious journal *Nature*, her first in the journal, in 1950. Dr. Franklin's expertise in X-ray diffraction technology was soon to become a key factor in influencing DNA history.

After finishing her post-doctorate training, in 1951, Dr. Franklin was invited to conduct new research at Kings College, London, England, to become a research associate in the department of physics and biophysics, under the direction of Prof. John T. Randall. In Randall's historical letter of appointment to Dr. Franklin, he had stated that, as far as X-ray diffraction experiments are concerned, Franklin and Raymond Gosling had their work to themselves. It later became a problem when Maurice Wilkins had believed that either Franklin was under his direction or that at the very least, they'd be working together on the same project of DNA. Franklin had rightly believed otherwise, of course. In any case, this episode started a long-lasting rift between Franklin and Wilkins.

Nevertheless, it was during this critical time that Franklin and Gosling had obtained their famous X-ray diffraction images, of which Wilkins was to share with Crick and Watson, leading to their deduction of the largely correct structure for DNA. It is unfortunate that in Watson's personal account of this historical era in his best-selling and famous book *The Double Helix*, his description of Franklin was a most unfavorable one. On the contrary, Franklin had been known to many of her friends, family and previous professional colleagues as a warm and amicable person. Watson has been soundly criticized for his portrayal of Franklin. It is clear that without Franklin's key data, it would have been terribly difficult for Watson and Crick to have accurately elucidated the DNA structure.

Giving up her future putative DNA work in March of 1953, Franklin moved to Birkbeck College, London, in the physics department, directed by John D. Bernal, to study the structure of the tobacco mosaic virus (TMV). Her X-ray diffraction work here was also extremely productive, showing that the RNA genome was located inside of the TMV proteins. This work was to become valuable to the field of virological research.

In 1956, Franklin was diagnosed with ovarian cancer and experienced an untimely death on April 16, in 1958, at the age of 37. It is widely speculated that had she not prematurely passed away Franklin would no doubt have been in the running for the Nobel due to her work on DNA. It has been further speculated that she might have very well been in the running for another Nobel for her work on the TMV structure. The rules of the Nobel commission state that the prize cannot be bestowed posthumously. After her death, Dr. Franklin has been widely recognized with numerous accolades, one being the building where she did her DNA work having been declared a national historic landmark. St. Paul's school opening up a Rosalind Franklin Technology Center, and the founding of the Rosalind Franklin Biotechnology Center at Delft, in the Netherlands.

James (Jim) Dewey Watson Jr. was born on the 6th of April in 1928 in Chicago, Illinois, USA, to parents of modest means, James D. Watson, Sr., his father, and Jean Mitchell Watson, his mother. Watson's sister, Elizabeth (Betty) was born 2 years later. As a child Watson was intensely curious, a quality he was to have throughout his lifetime. Growing up during the American depression, he was already recognized as having genius qualities and enjoyed a house filled with lots of books. The young Watson had an acute interest in birds and in books, and a prized possession of his has been a book, given to him in 1935 by his uncle, about the migration of birds. An early influence was his maternal grandmother, Elizabeth Mitchell ("Nana"), who lived with the family for a number of years.

In 1934, Watson started attending Horace Mann elementary school, where he was bullied—a situation that he grew to resent for many years to come. While in South Shore high school, Watson became interested in science after reading a 1925 Pulitzer-winning novel called "*Arrowsmith*" by Sinclair Lewis. Both Watson siblings finished high school early, at the age of 15, and attended the University of Chicago, both using academic scholarships and an early-admissions policy, to do so. Initially desiring to become a naturalist, Watson was immensely influenced by another book and a genetics professor.

The book was "*What is Life*" by atomic physicist Erwin Schrödinger; this inspiration was later to have a profound historical effect. One of Watson's college professors was Sewall Wright, a prominent geneticist who lectured about work from the laboratory of Oswald Avery in which DNA was thought to be the heredity material. It is at this point in his education that Watson became interested in the nature of the gene; and after he took his undergraduate degree in the field of zoology in 1947, Watson felt he needed to learn genetics. The four main take-home lessons he learned in college were 1) seek the original primary literature to acquire information 2) solve problems by figuring things out and formulating new theories 3) learn to think instead of memorizing facts and 4) seriously consider thinking about the most important things.

Rejected by Caltech and accepted by Harvard but with no funding, Watson thus entered graduate school in the Ph.D. program at Indiana University, in Bloomington, IN, USA, in 1947. Biographers of Watson relay that it is while in graduate school that Watson, perhaps influenced by others, developed an outspoken and brash manner, that he was to retain for the remainder of his life—sometimes placing him in rather dire consequences. Watson's graduate thesis advisor was Dr. Salvador Luria, a physicist by training who became a virologist and studied bacteriophages, which are viruses that infect bacteria. Watson was assigned the thesis project of studying the effects of X-rays on the bacteriophage infection upon their host bacteria. In graduate school, Watson was fortunate to spend his summers at Cold Spring Harbor, New York, and later Caltech, where he was greatly influenced by biophysicist, molecular geneticist, and one of the great pioneers of molecular biology, Max Delbrück. Watson eagerly became one of Delbrück's many protégés and consequently another member of the famous 'Phage Group' members of which were devoted to the study of the molecular genetics of the phage viruses. In May of 1950, Watson took his doctoral degree after writing up his thesis project dealing with phage inactivation and reactivation with X-rays. These results did not help Watson learn any more about the molecular nature of the gene, and he felt he needed to learn chemistry in order to do so.

A new postdoctoral student at the age 22 in 1950, Watson arrived in the research laboratory of Dr. Herman Kalckar in the University of Copenhagen, in Demark. Immediately after his arrival, he realized that he had made a huge mistake as he was bored having to learn biochemistry, instead of DNA. Watson was becoming convinced that DNA was the hereditary material, based on the work of Frederick Griffith and of Oswald Avery. Watson frequented another Copenhagen laboratory, where the research of which was directed by Dr. Ole Maaloe, and here Watson studied how radioactively-labeled phosphorous atoms were incorporated into progeny DNA, eventually publishing this work in the prestigious *Proceedings of the National Academy of Sciences*.

It was during this postdoctoral stint that Watson accompanied Kalckar on sabbatical to the Zoological Station, Naples, where Watson had attended a molecular biology conference and learned about the work of Dr. Maurice Wilkins, who presented his preliminary X-ray diffraction data on nucleic acids. It was about this time that Watson became aware of the great Linus Pauling and his then recent discovery of the alpha-helical nature of protein structures.

In 1951, Watson moved to Cambridge, England, to work at the Cavendish Laboratory, where he knew that DNA and other large molecules were being studied. It is here where Watson met Francis Crick, a graduate student working under Dr. Max Perutz. The story is told that about 30 minutes after meeting, Watson and Crick realized that they shared a common interest in DNA as being important for heredity and immediately became collaborators. Together, collaborators Watson and Crick put their brilliant minds together for the express purpose of solving the structure of DNA, before, they thought, the great Linus Pauling could do so.

Watson had visited Kings College to attend a seminar given by Rosalind Franklin. Watson, failing to take notes on the specifics of Franklin's X-ray diffraction data, incorrectly recalled the necessary parameters, which he and Crick then used to devise an incorrect triple-helical molecular model of DNA. It wasn't until Wilkins shared the requisite X-ray diffraction data, collected by Franklin, with Watson and Crick that they were able to come up with the proper correct double-helical model for the structure of the DNA molecule.

After publishing their proposed model for the DNA structure in 1953, Watson moved to Caltech and worked with Delbrück. Watson attempted to solve the RNA structure but failed. In 1955, Watson became an assistant professor in the biology department at Harvard, in Cambridge, MA. Meanwhile, data supporting the Watson - Crick Model for the DNA structure continued to pour into the scientific journals.

Sharing the Nobel Prize in 1962 in the area of Physiology or Medicine, Watson, Crick and Wilkins (for providing supporting evidence of the DNA structure) became widely celebrated for their historic and monumental achievement. It is considered by many scientists that the solving of the DNA structure is one of the greatest discoveries in scientific history.

During the Harvard years, between 1956 and 1976, Watson became a member of the National Academy of Sciences, wrote a definitive textbook "*Molecular Biology of the Gene*" in 1965, got married to Elizabeth Lewis in 1968, published his widely popular and famous "*The Double Helix*" in 1968, became director of the Cold Spring Harbor Laboratory (CSHL), also in 1968, and worked to devise policy on the safe use of recombinant DNA, in 1975. In 1976, Watson left Harvard and became full-time director of the CSHL, in New York.

Incidentally, Watson was to experience much trouble with his "*The Double Helix*" memoir. First, Crick (and others) became adamantly opposed to its publication because of the personal portrayal of himself and others. In fact, Harvard University Press had cancelled the project; although the manuscript was later accepted by Atheneum Press and by Weidenfeld & Nicolson. This was to create a rift between Crick and Watson, lasting many years. Second, Watson, and by extension, Crick, Wilkins and Perutz, were to be widely and enduringly criticized for their handling of the affair involving the incognito use of Franklin's X-ray diffraction data. Watson's personal portrayal of Franklin, in particular, was summarily and extensively criticized, as many of her close friends knew her to be an extremely intelligent, warm and friendly person.

In 1979, Watson published another popular textbook, *Molecular Biology of the Cell*, with Beatles album covers used as themes for the artwork in later editions. The current edition of the textbook is used by many institutions of higher learning in courses devoted to the study of fields that cover DNA.

In 1986, Watson became involved in establishing the Human Genome Project, a massive effort aimed at mapping all of the genes in the human genome and then determining the sequence of every nucleotide base pair in the human genome. Among the many donors of their DNA for the project, Watson's genome sequence was one of them; his full genome was published in *Nature* in 2007, using newly developed technology to complete it fairly quickly.

Starting in 2000, Watson began writing a series of essays, plus new memoirs and reminiscences, among them the sequel to "*The Double Helix*," called "*Genes, Girls and Gamow*" in 2002, plus "*DNA: the Secret of Life*" in 2003 and "*Avoid Boring People*" in 2007. In an unfortunate interview for a book tour promotion, Watson was quoted as making racist remarks. Even though he released a public apology, the damage was done. He was fired from some positions and resigned from others. His financial situation being dire, in an unprecedented move in 2014, he actually sold his Nobel medal at auction in order to conduct research—a first in history.

Francis H.C. Crick was born on the 8th of June in 1916 to middle-class parents Harry Crick and Anne Elizabeth Wilkins Crick, nearby to the town of Northampton, England. Harry Crick was in charge of a footwear production factory, namely of shoes and boots. The child Crick was described by his mother and Aunt Ethel (on his mother's side) as resembling an archbishop but with piercing eyes that implied a critical inquiry of the human subjects at hand. Like the young Franklin and Watson, Crick had also been recognized early on by others as being both extremely curious and gifted with exceptional intellectual abilities.

As a child, Crick was also an enthusiastic reader of various topics, including art, history and literature, but was primarily attracted to the subjects of science and math. At home, Crick set up a make-shift laboratory, attempting but failing to make silk, but succeeding in collecting flowers (earning first-place in the local science fair) and in producing small explosions—the latter of which was of a concern to his parents, of course. Interestingly, the story has been told that the young Crick lamented to his mother that by the time he grew up everything will have already been discovered, and there wouldn't be anything left for him later. His mother reassured him that there was indeed plenty left for him to discover.

Crick's first years of his formal education began at Northampton Grammar School, England. In 1930, Crick earned a scholarship to attend private school in London, at Mill Hill School. It is here that the young Crick eagerly read the entire freshmen chemistry textbook of Linus Pauling, "*General Chemistry*." He also studied medical biology, physics and, on his own, genetics.

In 1934, Crick enrolled at University College, in London, majoring in physics and minoring in mathematics. Crick took his Bachelors of Science undergraduate degree, with honors, in 1937. Crick had lamented that he was taught classical (historical) physics because the new modern physics topics, e.g., quantum electrodynamics, hadn't yet been discovered.

In 1937, Crick then entered graduate school in the Ph.D. program at University College studying water viscosity at boiling temperatures in sealed copper containers under the supervision of his graduate advisor Prof. Edward da Neville da Costa Andrade. Finding the project unfulfilling, Crick turned to the study of biophysics, but in 1939, World War II started, and Crick worked during the war as a civilian with the Admiralty, in the department of Mine Design; he also married Doreen Dodd and had a son, Michael, during this time. Returning to his Ph.D. laboratory at the end of the war, he found that his experimental apparatus had been destroyed during the war, freeing him from continuing with the water viscosity project.

With the war now over and at a loss about what to do with his career, Crick, in his autobiography, relates the advice given to him by his good friend, Georg Kreisel, a noted mathematician, who was reported to have encouragingly said to Crick, "I've known a lot of people more stupid than you who've made a success of it."

Crick had had interests in many topics but he soon realized that he was most interested in those topics in which he "gossiped" about the most to his colleagues, narrowing the topics to molecular biology and neuroscience. In 1947, Crick was accepted to once again continue his graduate career at Strangeways Laboratory, under the direction of Prof. Honor Bridget Fell, in Cambridge, England. During this period, Crick and Doreen divorced in 1947, and in 1949 he married Odile Speed. In his new laboratory setting, Crick cultured macrophages and studied the behavior of the insides of their cytoplasmic contents. His experimental efforts led to the publication of his first paper in a scientific journal, in 1950.

In 1949, Crick was accepted into the new research laboratory of Dr. Max Perutz at the Cavendish Laboratory, under the directorship of Sir Lawrence Bragg. The institution was established by the Medical Research Council, forming a unit for the purpose of studying the molecular structures of proteins. Thus, Crick set about studying the mathematics of X-ray diffraction upon protein structure, namely oxygenated hemoglobin, until, of course, the arrival of Jim Watson to the Cavendish Laboratory, in the fall of the year 1951.

Together, Crick and Watson set about to invoke the Pauling method of model building to eventually propose a largely correct molecular model for DNA that they had built using atoms built for molecular modelling. Again, Franklin's X-ray diffraction data being a key element in pushing Crick and Watson onto the correct DNA structure elucidation, they published their famous *Nature* paper in 1953. Interestingly, after they closely examined their structure, Crick and Watson immediately noticed that it pointed to a key element in its function to undergo replication—that the complementary base-paring produced a template function of one strand for the synthesis of a new strand of DNA, but Watson was plagued by the fear that perhaps they had gotten the structure wrong—so they left out this and other speculations about replications and other mechanisms in the *Nature* paper of 1953.

However, when new data were increasingly supporting their notion of the DNA structure, they felt emboldened to follow-up on their speculations by publishing these new ideas a little later. In the meantime, however, Crick got on with his protein structure work and completed his Ph.D. in 1954, in his late 30s.

Having just astutely participated in what is considered by many scientists and historians alike to be one of the greatest scientific discoveries of all time, the DNA structure, Crick then focused his attentions onto solving the genetic code. That is, it was not known at the time how the code embedded within the base sequences in DNA actually specified the amino acid sequences embedded in folded up and functional proteins.

Moving to the Polytechnic Institute in Brooklyn, NY, and collaborating with many other molecular biological converts, some of whom started the so-called RNA Tie Club, each of the 20 members receiving a tie with one of the amino acids on it, these investigators began the enormous effort required to solve the genetic code. In 1956, Crick formulated the so-called "adaptor" hypothesis, a molecule that is now what modern molecular biologists consider to be the tRNA molecule. These investigators also formulated their so-called Central Dogma about the flow of genetic information.

With much of the subsequent and compelling experimental data supporting the proposed Crick- Watson DNA structure, Wilkin's work amongst these data, Crick, Wilkins, and Watson became Nobel Laureates in 1962 in Physiology or Medicine, for their work. Working with Dr. Sydney Brenner, he and Crick formulated the term "codon" to describe the three-base code on mRNA that is read by the ribosome machinery to make protein. The codon principle of specifying amino acids is still used in all major textbooks that deal with DNA and the genetic code.

In 1977, Crick moved to La Jolla, CA, USA, to work at the Salk Institute for Biological Studies, where he switched fields completely to focus on neuroscience, namely, the study of consciousness. Another hobby, so to speak, was the notion of life on Earth being seeded by microbes from outer space, a phenomenon known as panspermia. Crick stayed on at Salk till his death on the 28th of July in 2004 at the age of 88.

5) On the television shows, we often hear about DNA matches. How important is our understanding of DNA to the average individual? Now how important is it for the average scientist?

I think that for the average individual, it is important to know about DNA for the following reasons. First, with the advent of rapid genome sequencing technology continually advancing for individual people and at increasingly inexpensive rates, individual people may now determine their entire personal genome sequences and consequently find out, for instance, that they may acquire a disease later in life, thus, having to make the necessary arrangements. Diagnosis of a disease may depend on the status of one's genetic composition. New medical treatments or novel useful products from biotechnology may be heavily dependent upon DNA. Therefore, one's general health status and quality of life may be significantly improved, if one understands one's DNA.

Additionally, people may disturbingly find out their parents are not necessarily whom they might have previously thought. Frequently, paternity disagreements require that the biological parents are definitively identified, and this in turn calls for knowledge of the DNA status for all parties involved. Forensic biological approaches that are conducted properly will often use DNA to help solve serious crimes, and to identify victims or their perpetrators. These DNA data are routinely presented in the form of a statistical value or chance that any DNA matches did not occur by a random chance.

Importantly, it will be in a person's best interest that one's genome composition and sequence data be kept private, so that, for example, those in the insurance sector don't obtain it, find out you may have (or might have at a later time) a condition perhaps requiring treatment, but that which might also be exploited by inordinately raising certain treatment costs and insurance rates. On the other hand, one's personal physician may find their patient's genome information handy in keeping them healthy.

For the average scientist, I'm convinced that it is imperative to know all about the biology and chemistry of DNA. The topic of DNA is central in basic courses in biology, genetics, cell biology, biochemistry, molecular biology, bioinformatics, immunology, physiology, microbiology, etc. As I mentioned earlier, just about every major discipline within the life sciences are going molecular. Without knowledge of DNA, it will become increasingly difficult for scholars to make novel and important discoveries in their chosen fields of interest.

Knowledge of the genomic statuses of entire populations of living organisms in all ecological and environmental niches may be tremendously useful in keeping our planet healthy for its inhabitants in the long run. Investigators have been interested in determining the biomes of various locations, such as the human gut, fresh waters, the Earth's vast oceans, and soil environments—these all involve collecting DNA samples, determining their genomic nucleotide sequences, comparing the retrieved sequences with all of the other sequences within gigantic sequence databases and learning the biology, the chemistry, the ecological relationships, etc. In particular, I think that our knowledge of the genome sequences of the microbes, many of which we already know, but many orders of magnitude more microbial genomes for which we don't, will become extraordinarily important for us all, and in many ways for which we have not yet anticipated.

6) What have I neglected to ask?

I feel extremely fortunate that someday I will be able to tell my grandchildren that I actually had the opportunity to listen to both Crick and Watson give scientific seminars. Being a biochemist by training I was converted to become a molecular biologist after reading Watson's "*The Double Helix*." After reading the book, I greatly wanted to know more about DNA. As a Ph.D. graduate student in a biochemistry and molecular biology program in the late 1980s and early 1990s, I had read the flyer on a bulletin board advertising Crick's presentation.

The seminar was, however, to be held at the Los Alamos National Laboratories (LANL), in Los Alamos, NM, about an hour-and-a-half drive through rugged mountains in my beat up old car. The trip made safely without car trouble and arriving on the night before Crick's seminar, I slept on the couch of my good friend and fellow graduate student, Kevin Laubscher. The next morning, the normally highly secure gates to the auditorium were now open to the public, for Crick's publicly held seminar. I followed Kevin as he moved past the gate entrance. I was, thus, greatly surprised as I approached the auditorium gate when the guard pointed to me still in line and asked what my nationality was—I replied that I was American (my family arrived in what is now New Mexico in 1598 with Oñate, some 20 years before the pilgrims arrived in Provincetown, but I felt this was simply just too much to explain that day).

Obviously ignoring that I said I was American, the gate guard then requested that he needed to see an ID, such as a passport, a visa, or a green card. I showed him my driver's license. After looking at the front and back he returned it and said nothing. Not knowing what to do, I brazenly walked through the gate. I never realized it would be so much trouble just to attend a seminar!

Anyway, it was all worth it because Crick's talk was fascinating! He took us through the story of the DNA structure determination, in technical detail—being a graduate student and having kept up with the literature, I was able to understand it, and it was all thoroughly enjoyable! I was captivated and mesmerized. I shall never forget that day.

You can perhaps imagine my good fortune, then, when during the following year, there was yet another flyer—this time advertising a seminar to be given by Jim Watson, in LANL, of course. This time, I made double sure that my beat up old car was tuned-up as I made my drive to the hill that is Los Alamos, NM. I slept on Kevin's couch again; the next morning, driver's license at the ready, I boldly flashed the license, said the obligatory "American" as I again brazenly flew past the gate entrance. In the LANL auditorium, I was not to be disappointed.

Watson took us through his technical version of the DNA talk, even going so far as to predict the good that shall come as a result of the human genome project. I didn't want the talk to end. It was so wonderful! And just when I thought the event was over, there was the announcement that students were invited to meet with Dr. Watson after the seminar, where there would be refreshments, as well.

Kevin and I made our way to the meeting room where we had hoped to meet with Watson and were supremely disappointed to find Watson there completely surround by non-students! We didn't know who these people were but they were most certainly not any students we knew.

Anyhow, Kevin and I literally elbowed our way into the front of the group, and there was Watson himself speaking, with everyone listening. We were only about a meter away from Watson—in the same room, and he was in the middle of irascibly criticizing Crick's notion of panspermia. In the midst of Watson's dressing down of Crick, I interrupted to ask the great Dr. Watson in front of everyone in the room what he had thought of "*What Mad Pursuit*?"

This is Crick's book in which he had given his account of DNA and of his involvement of the genetic code work, etc. I had read it just a few months prior to Watson's seminar. Watson looked directly at me for what seemed to be an uncomfortably long time and finally said "Oh that's a *good* book!" Then he went right back to the dressing down of Crick, precisely where he had left off, before I had interrupted him. Therein lies my personal account of Watson's account of Crick's personal account.

It was all too much! It was magnificent! It was great to know that these amazing scientists were also real human beings, just like ourselves. I shall never forget that day, either.

Chapter 31

Peter Mitchell

1) Peter Mitchell was the sole recipient of the Nobel Prize for Chemistry in 1978. Where was he born and where did he go to school during his early years?

Dr. Peter Dennis Mitchell, discoverer of chemiosmosis, was born in England, in a suburb of London called Mitcham, in Surrey County, in 1920 on the 29th day of September to parents Christopher Gibbs Mitchell and Kate Beatrice Dorothy Taplin.

Young Mitchell attended Streatham Grammar School starting in 1926, and in 1927, he moved to Barrow Hedges School, in Carshalton, a school he was reported to have hated, primarily because of bullying. Eleven-year old Mitchell then went to school in Taunton at Queen's College in 1931. It is during these years at Queen's in which Mitchell, while excelling in the sciences and mathematics, also developed an enduring interest in music, learning to play the violin. Unfortunately, Mitchell did poorly in geography and history, and he was disinterested in literature.

His main weakness at Queen's, however, turned out to be Latin, but not so much because he was not good at it; on the contrary, he was almost gifted in the discipline. He simply failed to enroll in the required course, which proved to be his undoing.

Despite his giftedness in Latin, Mitchell had no time to learn it, and he did poorly on the topic in his scholarship application examination to enter Cambridge. Luckily for Mitchell, the College's headmaster at Queen's, Christopher Wiseman, intervened by putting in a good word on Mitchell's behalf so that he could enter Jesus College, in Cambridge in 1939.

At Jesus College, Mitchell flourished. His interests widened in the arts, philosophy, and music, even learning to play the piano—his mother had given him a baby grand piano for his birthday. He also made lifelong friends, some of whom later played significant roles in his scientific career. A physiology faculty and Nobel Laureate, Prof. Edgar Douglas

Adrian, had an influential role in Mitchell's burgeoning interest in physiology, biochemistry and membranes.

In 1941, Mitchell majored in biochemistry and entered their biochemistry department at Jesus College in Cambridge. One interesting story is that one of his biochemistry professors, Malcolm Dixon, a young faculty at the time, lectured on respiration biochemistry by drawing enzymes as square boxes with arrows pointing to starting chemical groups (donors) in one part of the enzyme box and leading the arrows to the acceptor chemical groups in another part of the enzyme box. Although it was not Dixon's intent, it gave Mitchell the correct impression that within enzymes, these sorts of chemical groups moved around—an insight that Mitchell was later to exploit in his development of vectorial metabolism and especially in his ingenious elucidation of the mechanism called the Q cycle.

In 1942, Mitchell took his undergraduate biochemistry degree from Jesus College, Cambridge and entered graduate school in the same department at the institution. At first, Mitchell's graduate advisor was Dr. James Danielli, a membrane biochemist, but World War II interrupted Mitchell's thesis efforts. Mitchell participated in war-related work in which he developed chemical dyes to trace serum proteins to test serum leakage from cells exposed to the poisonous arsenic-laden gases, so that others could develop antidotes. The works were published and represented Mitchell's first forays into the scientific journals, two of which were published in the prestigious *Nature*.

Interestingly, while in graduate school at Cambridge, Mitchell was forever influenced in a most profound way by Dr. David Keilin, professor of biology and director of the Molteno Institute, also housed in Cambridge. According to some accounts, the story has it that at a small party, Mitchell and Keilin were holding a conversation and eating olives, when at some point during their conversation Keilin, without provocation or warning flung an olive across the room—to Mitchell's (and Keilin's) great delight! From that point on, Mitchell, too, took on Keilin's easy-going and relaxed but clearly anti-establishment attitude. I surmise this new outlook on life played a role in Mitchell's later efforts to establish successfully his own research institute at Glynn Manor at Bodmin.

Returning to his graduate thesis work after the war, in 1945, Mitchell learned that his thesis advisor left for another post, leaving Mitchell to acquire another thesis supervisor. Officially, Dr. Albert Chinball filled the vacancy, but in practice, at Chinball's request, Dr. Ernest Gale provided the requisite advisory function.

In 1948, Mitchell submitted his Ph.D. thesis write-up for consideration of acceptance by his committee. They rejected it, criticizing it for not being well focused and advising him to revise and resubmit later.

This first thesis had three main chapters to it. The first chapter was philosophical and physiological in nature and dealt with 'fluctoids' a term Mitchell derived on his own (he coined new terms throughout his life) which he meant to denote a connection between diffusion (fluctid) of molecules with their static distribution (statid). The second chapter dealt with bacterial cell wall surface molecules; and the third dealt with transport of the amino acid tryptophan into bacteria.

Instead of revising and resubmitting the rejected first thesis, Mitchell, being soundly chagrinned, took on a second altogether brand new project for a new Ph.D. thesis.

This second Ph.D. thesis project dealt with the mechanism of inhibitory action of penicillin upon bacteria. In it, he proposed that the antibiotic penicillin worked to kill bacteria by affecting their DNA synthesis activities. The committee accepted this second more-conventionally focused thesis, and Mitchell received his Ph.D. in biochemistry from Cambridge in 1951.

He even published several papers pertaining to this latter thesis work. Unfortunately, Mitchell's second Ph.D. thesis idea, accepted for his degree, that penicillin acted upon DNA, was later shown to be incorrect.

Penicillin's mode of action is at the level of bacterial cell wall synthesis inhibition, rather than at the level of DNA replication, as Mitchell had incorrectly proposed.

2) Mitchell first proposed the chemiosmotic theory in 1961. First of all, what is this theory all about and why is it important?

Mitchell's theory of chemiosmosis explains an important biological process that occurs during metabolism of foodstuffs to make biological energy. Briefly, Mitchell's Nobel Prize-winning theory illustrates how one form of biological energy is generated from nutrients and used to produce another form of biological energy, called ATP.

According to Mitchell, the first energy form is the proton gradient, while the second energy form is the more well-known ATP.

In more precise biological terms, the chemiosmotic theory states that the energy of the proton gradient (first energy form) established across a biological membrane during the respiratory chain oxidation-reduction transfer of electrons to oxygen is used as a driving force to make ATP (second energy form).

The chemiosmotic process involves coupling the respiratory chain machinery to the ATP-making machinery, in all living systems. Furthermore, the term oxidative phosphorylation describes how respiration coupled to ATP production leads to oxygen gaining electrons. As such, the 'oxidative' part refers to the loss of electrons from metabolic foodstuffs and their intermediates (electrons later are transferred to oxygen), while the 'phosphorylation' part refers to the addition of phosphate to ADP to make ATP. The overall process is as follows:

During nutrient utilization by living cells, the electrons stored in food are removed from glucose and other metabolites and then taken to the respiratory chain. As the electrons are shuttled from one respiratory component to another, protons are transported across the membrane. The protons build up on one side of the membrane creating a proton gradient. The energy of the proton gradient is used as the driving force to energize the ATP making process. In this manner, the respiratory chain is coupled to ATP synthesis by the proton gradient.

One example that illustrates the importance of the overall chemiosmotic process is that if one interrupts the respiratory chain, death of the individual will occur in minutes. The interruption process is known as uncoupling; that is, respiration uncoupled from ATP synthesis will signify a quick death, such as that seen with carbon monoxide or carbon dioxide poisoning. Thus, Mitchell's theory speaks to the heart of all life itself.

3) Apparently, Mitchell developed theories first and then did experiments later. Where exactly did he work over the course of his lifetime, and what were some of his main discoveries?

You are completely correct about Dr. Mitchell's propensity to come up with ideas before any supporting experiments were conducted. Mitchell performed most of his research work as a professor at Edinburgh University, in Scotland, and later at his own privately funded research institute called Glynn Research Laboratories, housed at Bodmin, in Cornwall, England.

In addition to the chemiosmotic theory, Mitchell was key to many other discoveries. A few of these discoveries are denoted below.

He envisaged the idea of the so-called 'vectorial metabolism' to describe the movement of protons (H^+) and hydroxyl ions (OH^-) across the membrane. The idea was later expanded upon to include the transport of virtually all other biomolecules across the biological membrane.

Further, Mitchell coined the term primary active transport to describe the concomitant processes of a catalytic chemical reaction and the transport of molecules across the membrane. Nowadays, the term primary active transport has been refined to describe the use of ATP energy to perform transport against a solute concentration gradient.

Importantly, Mitchell came up with the term proton-motive force (PMF), which is the energy stored in the proton gradient and which is used to drive the synthesis of ATP. Early investigators in Mitchell's time did not believe in the PMF, and these contemporaries would publically ridicule him, referring to the PMF as the 'Peter Mitchell's Fallacy.' Mitchell had been correct.

Mitchell coined the term 'symport' to denote transport of solute across the membrane as driven by an energetic ion gradient. The process has also been called 'co-transport' and has been distinguished from 'monoport' (later uniport), which is the transport of one solute across the membrane, and antiport (transport of solute and ion across the membrane in opposite directions).

Mitchell discovered the phenomenon of 'uncoupling' the respiratory chain from ATP synthesis using a chemical called 2,4-dinitrophenol, which would function to dissipate the proton gradient. The result was the collapsing of the energy stored in the PMF, which then halted ATP synthesis.

Mitchell had discovered that the energy coupling between respiration and ATP production was mediated by ion gradients. While studying these energy coupling systems, he also found that the membrane itself formed a barrier and did not itself play a role in proton translocation across the membrane. Complexes within the respiratory chain and the ATP synthesis machinery itself, called ATP synthase, instead performed this proton transport across the biological membrane.

Mitchell invoked the term 'ligand conduction' to describe proton translocation across the membrane through the respiratory chain complexes and the ATP synthase molecule.

Importantly, Mitchell formulated a clever scheme, called the Q cycle, while experiencing a bout of insomnia during the night of May 20, 1975. The Q cycle explained how protons and electrons move through one of the respiratory chain components, called complex III, or the ubiquinone-cytochrome C oxidoreductase.

Mitchell invoked the term 'proton well' to illustrate how the protons that built-up on one side of the membrane would collect within the ATP synthase molecule.

Taken together, these scientific contributions by Mitchell make him a key pioneer, if not one of the founding fathers, of the important field of bioenergetics.

4) Mitchell seemed to study the link of metabolism with membrane transport. Why is this important and why is it relevant?

In a sense, metabolic machinery first works by garnering electrons stored in food molecules, like glucose, fatty acids, nucleic acids and amino acids. These electrons are then concentrated together and taken to the respiratory chain, which resides in the inner membrane of mitochondria from all animals (including all humans), in the thylakoid membrane of chloroplasts from plants, and in the plasma membrane from all bacteria.

Mitchell linked metabolism with membrane transport in three distinct ways. First, he connected the metabolic electrons and their transport along the respiratory chains with proton transport across the various membranes, to produce the necessary proton gradients, a new form of biological energy that's used to make ATP.

Second, Mitchell connected the reverse translocation of protons across the membrane with the synthesis of ATP. During transport of protons across the membrane down their concentration gradient, i.e., from a high proton concentration on one side to a relatively low proton concentration on the other side of the membrane, ATP is consequently formed.

Third, Mitchell's work connected nutrient uptake that's performed by living cells to their own metabolic purposes. All living biological cells allow entry of nutrients into their cytoplasmic spaces to be metabolized for energy production. The energy produced allows the cell to carry out their various biological functions.

These three aspects were shown to be true for animals, plants and bacteria alike. This work is important in that it speaks to the energetics inherent in all living organisms, all of which require energy for life. Thus, without this needed energy, there would essentially be no life.

The ATP molecules made by living beings are necessary to fuel the processes required to make cells, tissues, organs, and to conduct the physiological, biochemical and cellular systems, all of which are important for all life on Earth.

5) Apparently, Mitchell did not like working in higher-ed, so he started his own research institute called "Glynn." What discoveries came out of Glynn?

The story of Mitchell's establishment of Glynn Research Laboratories is steeped in legend. Glynn was an old manor house converted into an independent and privately funded research institute, headed by Mitchell.

One legend posits that Mitchell was disillusioned by academic life, quit his tenured professorship at Edinburgh, started a lucrative dairy farm business with his brother, and bought a castle in which to live and conduct research. While this story is great, it is inaccurate.

It is more accurate to say that Mitchell's health caused him to take an extended leave of absence from Edinburgh. The dairy business came later. Funding for Glynn was provided partly by an inheritance to Mitchell and by benefactors. The 'castle' was actually an old dilapidated manor that was infested with fungi and which he had originally meant to restore as a summer home, but later became the Glynn Institute.

His ultimate decision to resign from the university post, however, was indeed influenced by his anti-establishment sentiments (the 'Keilin effect'), but it was also the result of his constant fight with the academic and research establishment, both of which had staunch opponents to his chemiosmotic theory.

These latter incidents have been referred to as the so-called 'Ox Phos Wars' a direct reference to Mitchell's alternative views regarding how ATP production was energized during oxidative phosphorylation (Ox Phos). Mitchell had correctly proposed that the proton gradient was the energy to make ATP while many in the establishment held that an elusive 'high energy-bond intermediate metabolite' was the necessary energy for ATP synthesis—none was ever found. While such a high energy-bond intermediate, ATP, was most certainly true for intermediary metabolism, as rightly proposed by Fritz Lippmann, it was just as equally NOT to be true for the making of ATP.

Mitchell's biggest discovery at Glynn was the so-called Q cycle. In his Q cycle hypothesis, Mitchell addressed the nature of the movements of protons and electrons through the ubiquinone-cytochrome C oxidoreductase system, known also as complex III, present in the respiratory chain. The Q cycle was a clever and insightful contribution to science. The Q cycle involves four main steps, each of which trace the movement of electrons and protons within Complex III. Mitchell's Q cycle has a prominent place in all textbooks dealing with nutrition, metabolism, biochemistry, animal and microbial physiology, and energetics.

6) According to many accounts, he was quite a character, and quite a ladies' man. Do you have any great stories about his amorous exploits?

In university, while a student, accounts state that Mitchell was considered attractive by many an admirer, a flamboyant dresser, and was described as having had a flowing mane of hair to his shoulders, reminiscent of that of a young Beethoven. He seemed to project a lasting presence among potential female devotees.

According to biographers of Mitchell, starting in college and well into his later years, he apparently had a propensity for being overly keen with girlfriends and spouses of his friends and colleagues, alike. Furthermore, many of these significant others of others were apparently equally as taken with Mitchell. Many found him both charming and witty. One of these attractions involved a spouse of a fellow biochemist. Both of them left their own spouses and for a brief while moved in together in a bungalow nearby to Glynn. The affair did not last long (a week), and each returned penitent, back to their respective spouses. The affair had placed a lasting strain on the marriage between Mitchell and his then second wife, Helen.

One seemingly amusing anecdote involves one occasion at a wedding. During the gathering Mitchell noticed an attractive female from across the table, whereupon he inquired of her whether they had ever met before, to which she replied that indeed they had; she introduced herself as his first former wife, Eileen.

7) At the end of his life, Mitchell sadly had a number of health and medical problems. What do we know about these?

Mitchell suffered from recurrent stomach ulcers. At the time, such ulcers were considered the result of stress and type of lifestyle. As alluded to above, one of these ulcers was so serious that Mitchell had been advised by physicians to rest and relax to minimize the stress—remember, he was in the midst of the Ox Phos Wars. The extended leave of absence from Edinburgh due to his ulcers was one of the prime factors that eventually led Mitchell to resign his academic post and start up his work at Glynn. It was not until the late 1970s that one of the true causes of stomach ulcers was found to be a bacterial agent called *Helicobacter pylori*. Nowadays, such infectious maladies may be effectively treated with antimicrobial agents.

Shortly after the fiasco of the extramarital affair mentioned above in 1977 Mitchell suffered a nervous breakdown. It represented a turning point in Mitchell's life. He reorganized the administrative structure at Glynn. Interestingly, his creative genius for generating new ideas diminished and gave way to create new and ingenious ways to defend his largely correct earlier ideas from the attacks of his numerous critics.

Sadly, Mitchell, a supreme aficionado of classical music, who played the works of Mozart, Bach, and Beethoven on the piano, lost his hearing. He had developed an ear infection, which required surgery. Unfortunately for Mitchell, the operation was botched, causing him to be deaf for the remainder of his life. It affected him greatly as music had been a refuge.

8) What have I neglected to ask?

Of the various reasons for the overall scientific establishment actively opposing Mitchell's chemiosmotic theory, one of these being his having gone against the prevailing notion of the high energy bond intermediates, there was another prime reason: his published journal articles were exceedingly difficult to read.

He wrote in relatively long and convoluted sentences, and he developed his own nomenclature. Some of his nomenclature was eventually adopted (e.g., vectorial metabolism and proton-motive force) but many other new words (e.g., proticity, chemicalicity) that he coined failed to endure the test of time.

Ernesto Uribe and Andre Jagendorf conducted one of the first published studies that seemed to turn the tide towards eventual acceptance of Mitchell's version of oxidative phosphorylation in 1966. In their experiment, Uribe and Jagendorf added an acidic solution to chloroplasts (which harbor all of the necessary components necessary for chemiosmosis to work) to equilibrate the protons on both sides of the membrane. Next, they added an alkaline solution to the equilibrated chloroplasts to establish an artificial proton gradient. Lastly, they added ADP plus radioactive phosphorous-32 (^{32}P) and observed a tremendous burst of radioactive ATP production!

Interestingly, one of his most vocal opponents, Dr. Efraim Racker, had later published key evidence in 1973 that actually supported Mitchell's chemiosmotic theory. Racker and co-author Walter Stoeckenius tested the prediction based on Mitchell's idea regarding oxidative phosphorylation that the proton gradient was connected to the respiratory chain to provide energy for ATP synthesis. In their experiment, they mixed tiny membrane vesicles containing a purified light-driven proton transporter from bacteria with purified ATP synthase molecules taken from bovine heart muscle mitochondria. Adding light to turn on the proton pumps they invoked the production of a proton gradient. They watched as the protons went through the ATP synthase, making ATP on the other side of the membrane vesicles!

Many new experiments in the field were subsequently published, and all or most of these supported Mitchell's version of bioenergetics. Hence, many of the detractors of Mitchell's chemiosmotic theory eventually came around and actually became some of his strongest proponents. They know who they are.

Chapter 32

FREDERICK SANGER

1) Dr. Varela, we first have to mourn the passing of Frederick Sanger who recently passed. Obviously he was born in Great Britain, but what do we know about his early life and childhood?

A laureate of two Nobel Prizes, Dr. Frederick (Fred) Sanger was born on the 13th of August in the year 1918 to parents Frederick Sanger and Cicely Crewdson Sanger in the small country village of Rendcomb, located in the county of Gloucestershire, the Cotswold district, in England. Sanger's father was an established physician in the small community, and his Quaker mother was a homemaker who kept a diary detailing events concerning mainly about the household and her family with the three children, young Fred being the middle child. Fred had an older sibling, Theodore (b. 1917) and a younger sister, Mary (b. 1923). The Sanger children had a governess, and the family environment is reported to have been a positive, warm, and friendly household. In her diary, Cicely recorded insight into the behavioral temperament of young Sanger. Sadly, however, she did not have the opportunity to record anything beyond his undergraduate years because she passed away in 1938. His father had died even earlier, in 1937.

In 1927, nine-year old Sanger was enrolled in a Quaker preparatory boarding school called The Downs, located in Malvern. Although the experience is reported as being a traumatic one for Sanger, as he was bullied, he was, nevertheless, to earn outstanding grades while enrolled at The Downs.

Soon afterwards, in 1932, 14-year old Sanger was enrolled in another newly established boarding public school (in England public schools are actually private schools as a fee is assessed) called Bryanston School, in Dorset, England. Although Sanger failed to secure a lucrative scholarship to attend, he was nonetheless quite happy to be a student at this particular institution where he flourished, especially in the subjects of biology and chemistry. In fact, Bryanston was such a favorite school that many years later (in 2007)

Sanger was to establish a new building in his name, the Sanger Centre for Science and Mathematics, at the school.

In 1936, Sanger attended St. John's College, Cambridge, his father's alma mater, majoring in the Natural Sciences and emphasizing in Biochemistry. During his college years, he became a conscientious objector and first met his future wife, Joan Howe.

Initially hoping to ultimately obtain an M.D. degree (perhaps it was more a hope entertained by his father, rather than by Sanger) the death of both of his parents while still in college made Sanger feel that his chances for advancement in medicine were somewhat compromised. Thus, it is during this time-period that Sanger chose instead to pursue scientific research as a full-fledged career. In 1939, he took his B.A. degree from St. John's College.

2) His first Nobel Prize was for the sequence of protein. Why is this important in the big scheme of things? When was that awarded, and what did we learn from this?

Sanger earned his first of two Nobel Prizes in 1958, in the area of Chemistry, for his investigative work with the important hormone called insulin. He was the sole recipient of the 1958 award in the category.

Sanger was the very first scientific investigator in the world to elucidate the entire sequences of the amino acids along the two protein chains that make up insulin. Solving its protein sequence was such an important finding because knowledge of the nature of the amino acids along the insulin protein chains, or of any protein, almost immediately allows an investigator to learn about how the protein chains fold into their three-dimensional structures. Along these lines, knowledge of the overall 3-D shapes of biologically and biochemically important molecules lends tremendous insights into the nature of their mechanistic functions. All of these great advancements in protein biochemistry started with Sanger's study of insulin.

Sanger studied a series of biochemical reactions to help him elucidate the nature of the amino acid sequence of both insulin protein chains, called alpha (or A) chain and beta (or B) chain.

3) Apparently, twenty-two years later, the Nobel Committee awarded him the 1980 prize in Chemistry once again for finding a way to determine the ordered sequence of DNA molecules. I suspect this has something to do with the twisted ladder effect? Or am I off on this?

On the contrary, you are not off on your assessment of Sanger's important work with the double helical DNA molecule. Sanger's studies on DNA earned him his virtually unprecedented second Nobel Prize, also in Chemistry, in 1980. As of this writing, only four other individuals in history have garnered 2 Nobel Prizes, and Sanger is apparently the only person ever to have gotten two Nobel Prizes in the Chemistry category.

Sanger was able to determine the nucleotide base sequences of DNA by again putting his knowledge of biochemistry to work. He used bacterial enzymes like, for instance, DNA polymerase which he obtained from the bacterium called Escherichia coli, and other chemicals to break up DNA into nested pieces. Each of these DNA pieces differed chemically by one base-pair in their lengths. Next, he collected his various pieces of DNA and placed them in so-called electrophoretic gels, ran an electrical current through the DNA-laden gels and then read the radioactively labeled bands on the gels on exposed X-ray film using a light box. From this rather labor-intensive work he was able to determine the entire genome sequence of a bacteriophage virus called ΦX174 (the Greek symbol Φ is pronounced phi).

Application of Sanger's method for determining the nucleotide base sequences along their twisted DNA chains was a remarkable and profound discovery.

4) Proteins seem to have a very exact, defined, specific sequence or composition. How was Sanger involved in this and why is it important?

Protein chemists, biochemists, and molecular biologists often like to proclaim that although DNA is indeed universally considered to be the "molecule of life," it is the proteins, however, which "do everything." Proteins are molecules that partake in the performance of those sort of processes involved in all living systems. Each protein has a specific function or set of functions, and these functions are carried out by the particular shapes, or molecular configurations, inherent within each of the proteins.

These protein structures, in turn, are dictated by the characteristic sequence of amino acids within them. It can be said that proteins are actually chains of varying lengths which are composed of amino acids that are tied together by peptide bonds. The amino acid chains of polypeptides in turn will fold up around each other to form molecular shapes that are virtually limitless in their structures formed but are certainly specific to each and every distinct protein. Thus, the particular sequences of amino acids along protein chains can be said to determine the precise molecular natures of protein structures. Therefore, if one knows the molecular shape of a protein one might then know the biological function of those proteins.

Sanger's involvement with this protein chemistry field comes into play as he was the very first scientific investigator to determine the sequence of amino acids for a protein, namely, that of insulin.

Let's briefly consider Sanger's Nobel Prize winning work with protein sequence determination of amino acid residues along the insulin hormone protein.

First, Sanger broke up each of the two insulin A and B chains from each other and then broke apart each of these A and B polypeptides into smaller fragments. He had denatured the two chains from each other by adding chemicals to undo the disulfide bonds connecting the A and B polypeptides to separate them from each other and then, using a technique called chromatography, he isolated each of the separated chains.

Then, he developed a 'labeling' molecule that would light up or fluoresce when induced and that would bind to the one amino acid that was located at one of the ends of a separated insulin protein chain, whether it was the A or the B chain. The labelling molecule was called 1-fluoro-2-,4-dinitrobenzene (FDNB).

Next, he used very strong acids to disrupt each of the two insulin protein chains, and the amino acids would come apart from each other releasing them including the end-labeled amino acid. He did this disruption for each of the A and B chains separately.

Fourth, Sanger invoked a special biochemical process called the 'Edman degradation' which had been developed by the biochemist Dr. Pehr Edman. The process also end-labels protein chains, like Sanger's FDNB, but instead when the protein chain degrades, only the end-labeled amino acid comes off, not all of them like in Sanger's strong acid treatment.

Next, Sanger would conduct a series of chemical and biochemical reactions meant to determine the identity of the amino acid that came off of the individual chain during the earlier third step.

Lastly, Sanger then purified and re-used the insulin protein that was left over from the third step with its missing (though identified) amino acid, but essentially intact otherwise, to start another round of the Edman degradation. Then he would simply start over again, label the protein, remove the end-labeled amino acid with the Edman step, and identify the next amino acid along the chain till he was finished with identifying all of the amino acids in the protein.

Now, with insulin consisting of two separate polypeptide chains, the A and the B chains, and each of the polypeptide chains being unique, sequence-wise, it was a very labor intensive process to identify all amino acids in the protein. Insulin's A chain has 21 amino acids, and the B chain has 30 amino acids.

It took years to complete the entire protein sequence identification for both insulin chains. Much later, the cumbersome protein sequencing technology became automated, the new advanced machines being given the name 'sequenators.'

The B chain gave away its amino acid sequence secret first, and its sequence was published in 1951, with co-author, Hans Tuppy. The A chain gave away its sequence secret a couple of years later in 1953. No one had ever accomplished such a remarkable task before in scientific and world history.

5) Key word – electrophoresis - what does this mean and why is it relevant?

The term refers to a laboratory technique that's very often used to separate biological components from each other in order to study each of the components separately. The electrophoresis process involves making a 'gel' often in a rectangular shape, like that of a flat brick, but with the consistency of Jell-O. In fact, investigators making electrophoretic gels often use the same sort of materials in Jell-O, like gelatin, or agarose, or perhaps more sophisticated materials, like polyacrylamide, depending on the purpose of the experiment and the nature of the biological molecules being studied. The gels will also have little 'wells' in them, formed during the gel making process. The wells will be located on one end of the gel brick and used as little buckets to place the biological samples.

Once an electrophoretic gel is made, complete with its wells, the investigator will then load the wells with the molecules in question, like proteins, nucleic acids, etc.

Next, here's where the 'electro' part comes into play. The loaded gel will be placed in a buffer with ions and subjected to an electrical current. The current will force the samples loaded in the wells to move into the insides of the gel, called a gel matrix, and redistribute themselves according to their relative sizes and charges. The smaller molecules will move faster and farther along the gel than their larger counterparts. The more negatively-charged molecules will move 'faster' and farther along the gel than their positively-charged relatives, which stay somewhat behind. Sanger was known to have used electrophoresis when he developed his Sanger method for DNA sequencing.

6) The "Sanger method" supposedly allowed very long stretches of DNA to be accurately and rapidly sequenced. Why is this important in the big scheme of things?

The Sanger method of DNA sequencing has also been referred to as the 'Sanger dideoxy' or 'chain termination' method. This Nobel-winning work involved the following steps.

First, Sanger added the basic ingredients, such as the DNA to be sequenced, a short DNA primer to start DNA synthesis, a DNA-making enzyme called DNA polymerase, an excess amount of the nucleotide bases, G, A, T, and C, plus a tiny amount of Sanger's special DNA chain terminator molecules called dideoxy-G, dideoxy-A, dideoxy-T and dideoxy-C. These chain terminators, usually labeled with radiation or a florescent probe, will prevent the continued synthesis of new DNA chains from that point onward.

During the DNA-making reaction, sometimes the machinery added incorporates a regular base into the growing DNA chain, or it incorporates a chain terminator, instead. The reaction proceeds in which new DNA is made but because of the chain terminators, the end-product is a series of new DNA chains of varying lengths, each usually differing by only one nucleotide, if performed correctly.

The mixture of new DNA of different lengths was placed by Sanger into wells of an electrophoretic gel and subjected to an electrical current to separate the fragments according to their lengths, from larger to smaller. The gels are placed on photographic film to expose the DNA, and the nucleotide bases are 'read' along the length of the chains, thus revealing the DNA base sequences.

The DNA sequences are stored and archived in computer databases. The DNA sequencing process is important for a variety of reasons. The knowledge of DNA sequences, for instance, facilitated the cloning of genes and for computer determination of their corresponding protein sequences. This alleviated the need for cumbersome protein purification and protein sequencing efforts, saving a tremendous amount of time and money.

Also important was the ability of investigators to study sequence variants of DNA genes and proteins, such as in genetic diseases, or their evolution, or for garnering more efficient industrial efforts to make new products, such as medicines.

7) He studied lysine and insulin. What discoveries or contributions did he make relative to these two things?

While an undergraduate student, and to Sanger's great surprise, he excelled in Biochemistry, earning the highest honors possible and, thus, qualifying him for advanced studies in graduate school. Sanger's work with lysine began in graduate school.

Thus, in 1940, Sanger was granted admission to graduate school at Cambridge, entering the laboratory of Prof. Norman (Bill) Pirie, housed in the institution's Biochemistry department. In 1936, Prof. Pirie had discovered that a virus, called the tomato bushy stunt virus, was amenable to crystallization. In Pirie's laboratory, Sanger was given the thesis project involving the purification of an edible protein from grass. However, very shortly after starting on the new thesis project, Pirie left the institution for another academic post, leaving Sanger to find another thesis advisor at Cambridge.

Consequently, the new graduate advisor whom Sanger chose to work with was a newly appointed faculty member by the name of Albert Neuberger. Many years later, Sanger would recount how it was Prof. Neuberger himself who taught him (Sanger) how to properly conduct scientific research and, importantly, not to be afraid of studying new things and to not despair too terribly if an experiment did not work out—just try another experimental route.

In Neuberger's laboratory Sanger set out to study the degradative metabolism of an important and 'essential' amino acid called lysine, in laboratory rats. Sanger also partook in so-called war-work pertaining to World War II, namely, the analysis of nitrogen and protein composition in different strains of potatoes.

Although Sanger was unable to definitively elucidate the entire degradative biochemical pathway for the metabolism of lysine, he was nevertheless able to advance the work sufficiently enough to publish five papers on the topic and in his potato work to finish his Ph.D. thesis project, and make a very good starting point for other investigators to complete the lysine breakdown pathway. Sanger defended his Ph.D. thesis project in 1943 and took his doctorate in Biochemistry from the University of Cambridge.

Sanger's insulin work involved elucidating the entire amino acid sequences of the two polypeptide chains, called A and B, that make up the hormone insulin. He used labor-intensive and harsh chemical techniques in order to do so. He had worked with the insulin molecule from bovine, or cows. The novel and Nobel Prize worthy insulin sequence work ultimately led to elucidation of the sequence for human insulin. It then led to the cloning of the human gene encoding insulin; and it further led to the determination of its three-dimensional protein structure by Prof. Dorothy Crowfoot Hodgkin, the 1964 Nobel Laureate in Chemistry.

8) Insulin, is obviously tied to diabetes- a major medical problem- but what exactly is lysine, what does it do and why is it important?

Lysine is essential in the sense that humans and other animals cannot make this particular amino acid and must, therefore, acquire it from the diet. Good sources of this nutrient include foods rich in protein like meats, fish, legumes, poultry, etc. This amino acid is used by the body to make larger protein molecules, which we know are vitally important for life's processes.

Much biochemical work has been placed on elucidating its metabolism. The biosynthetic pathway for making lysine has been well defined. Likewise the catabolic pathway for degrading the amino acid into its smaller chemical parts has been definitively determined. Both aspects of its metabolism are regularly included in biochemistry and physiology textbooks. The topics are routinely taught to students of healthcare sciences, like medicine, dentistry, and in the biomedical sciences.

It's been postulated that lysine is important for immune system modulation and as an anti-anxiety supplement. It has been used as an external dietary supplement, a characteristic featured in a popular television show, 'The Big Bang Theory' about science and computer 'nerds' in which one of the main characters touts the virtues of lysine. Its use is also important in agriculture for feeding of farm animals. Industrial production of lysine involves the fermentative process using microbes and various sugars as a starting point for lysine synthesis. Recombinant DNA technology, made possible by Sanger's DNA work, has been used to boost the production process for lysine.

It is interesting to note that in Michael Crichton's book '*Jurassic Park*' the fictional book in which dinosaurs are brought back to life using ancient DNA fossilized in amber samples of mosquito blood meals, the cloned dinosaur DNA was genetically engineered to be defective in their ability to make lysine. This made the cloned dinosaurs dependent upon the fictional scientists for their source of dinosaur dietary lysine.

9) He apparently was called "the father of the genomic era." Can you explain this for us?

Because of Sanger's DNA sequencing invention, it allowed the elucidation of the molecular nature of both genes and proteins. Furthermore, improvements in the efficiency of Sanger's work was greatly facilitated by converting from the slow and cumbersome gel electrophoresis to the more advanced and efficient capillary approach. New technological improvements, such as the capillary mode, allowed investigators to sequence DNA with greater efficiency and more rapidly than ever anticipated. It became possible to map genes and to sequence the DNA of entire genomes of organisms, thus, instituting the new field dedicated to the study of whole genomes from living organisms. The new discipline is referred to as genomics. Sanger's DNA work made the new and burgeoning genomics field acutely possible.

Prior to the improvements in gene sequencing technology in the 1980s, it had been a laborious task to sequence even one gene. Entire Ph.D. theses might have been devoted solely to the cloning and sequencing of a particular gene of biological or biomedical importance.

With the advent of genomics, it became possible to determine the entire nucleotide base sequence of an intact (i.e., whole or complete) genome. Furthermore, it was possible to compare genome sequences with other genomes from other individuals or other species. This comparative genomics approach has made possible the study of the evolution of living organisms as they interact with other species or how they behave in their native or novel environments. Such evolutionary information may shed light into what the future has in store for generations of living beings throughout the world.

Comparison of genomes has greatly enhanced the search for targets with novel biological mechanisms and therapies. For example, in our research laboratory, we determined the entire genome sequence of a non-pathogenic variant of a dangerous bacterium called *Vibrio cholerae*, the causative agent of cholera—a serious disease in many parts of the world. Using our genome sequence data, we then compared our 'good' bacterium with the known genome sequence of the classically dangerous cholera bacterium. Identifying the differences between the good and not-so-good bacteria allowed us to recognize those genes in the cholera-causing bacterium which might be involved in conferring the serious nature of its effects on their patients. Thus, these new genes may encode cellular targets that we may be able to exploit by reducing their activities in order to minimize their deleterious effects.

Currently, it is possible to take samples of DNA from an environment and ascertain the entire organismal content of that environment by sequencing and learning about the entire make-up of all of the species that reside in that environment. This has been done recently with the human gut, which is predicted to harbor thousands, if not millions of microbial species. The gut microbiome has been implicated in determining various aspects of human health, such as immunity, behavior, and personality, such as whether one is generally a happy individual. Much work will be needed to determine the extent to which each of the millions of bacterial species interact with others. In fact, with genomics-based studies, it has been recently postulated that Earth harbors a trillion microbial species. If true, it will require efforts based on genomics to find and identify all of these yet undiscovered species.

10) There are all kinds of stories about him- apparently he was a Quaker, a pacifist, and, I quote "A chap who just messed about in his lab." What other gems do you know about this iconic scientist?

It is interesting to note that despite his advancing age, he spent a great deal of time literally 'at the laboratory bench' doing experiments with his own hands up until the time of his retirement. Many senior scientists are often loathe to work at the bench or are encumbered to do so by having to perform so many 'required' administrative duties, or to participate in grant-writing and manuscript-preparation, or serving on numerous committees, etc.

Prof. Sanger was known to be a great scientific mentor, with several of his doctoral and post-doctoral students themselves having earned a Nobel of their own. One of his Nobel mentees participated in structural studies of antibodies while another was involved in studying the molecular edges of chromosomes, called telomeres.

When asked in an interview how he felt about being the recipient of two Nobel Prizes, he replied that it was more difficult to get the first one than it had been to get the second one. The first Nobel made it possible to more easily acquire new laboratory facilities and other research opportunities that made his subsequent research endeavors more readily possible.

It may interest our readers to know that Sanger's laboratory was not the only one devoted to discovering how to sequence DNA, a procedure that was highly sought after by many molecular biologists of the day. The Harvard laboratory of Walter Gilbert proved to be Sanger's biggest competitor. Gilbert's approach was called the Maxam (after Allan Maxam) and Gilbert chemical method. The story is told that once during a Gordon Research Conference, members of Gilbert's lab presented their sequencing methods and data without referencing the work of Sanger, which they (members of the Gilbert lab) had used. This slight had rankled Sanger. Nevertheless, Gilbert and Sanger shared the 1980 Nobel in Chemistry for their efforts in inventing the DNA sequencing technology.

When Sanger determined the genome sequence of the ΦX174 phage, he found that there were two overlapping genes encoded within the genome. Thus, Sanger was one of the first investigators to go beyond the 'one gene, one enzyme' notion that was prevalent throughout the field of molecular biology.

Another finding by Sanger that was of great importance was his sequencing of mitochondrial DNA. The mitochondria are sub-cellular organelles that function to produce the cellular energy needed for life and are thought to have arisen from an ancient bacterial infection of an ancient eukaryote. When Sanger sequenced mitochondrial DNA, he found that the genetic code was not entirely universal, as was once thought.

Others who knew Sanger well on a personal basis have observed that he was quiet, humble, humorous, and often partook in self-deprecating behavior. It has been reported that he refused the offer to be knighted by the Queen, remarking only that he had not wanted to be changed by the honor and to be called 'Sir Fred.'

Chapter 33

ROSALYN YALOW

1) Rosalyn Yalow is one of the most outstanding females in science, and medicine. And she won the Nobel Prize for medicine. What do we know about her early days, and parents and upbringing?

Indeed, Dr. Rosalyn Yalow, Nobel Laureate and one of the greatest scientific investigators of all time, directly changed the course of history because of her efforts spent as inventor and discoverer of the so-called radioimmunoassay (RIA), a technique that was used to make significant advances in many scientific fields, including physiology, immunology, biochemistry, endocrinology, cell biology, and cancer. Dr. Yalow was one of the first American women to earn the Nobel, in the fields of medicine or physiology, in 1977.

Rosalyn Sussman Yalow was born on the 19th of July in the year 1921 to Jewish parents Simon Sussman and Clara Zipper Sussman in New York City, NY, USA. With one older brother, Alexander (b. circa 1915), the two Sussman children grew up in the south Bronx district, a working-class neighborhood, and spent their early childhood in poverty—the latter of which was primarily due to the stock market crash of 1929 and the ensuing Great Depression of the 1930s. Her father was a merchant of neckties, paper, and twine in a New York district filled mainly with an immigrant population. Both children worked in the family business, with young Rosalyn cutting out necktie patterns and turning out collars that her mother had sewn together.

Simon was of Russian descent living on the lower eastside of New York and with an 8th grade education, owing to his family's poverty and having to work. Rosalyn's mother Clara was a German immigrant and was educated up to the 6th grade, a consequence of constantly having to relocate from place to place. Simon was not a book reader but did keep up with the daily newspaper events. On the other hand, Clara read her children's schoolbooks and consequently learned fluent English and other subjects on her own. Clara even managed to read her children's university-level textbooks.

The child Rosalyn Sussman already showed signs of brilliance; she had learned to read even before being enrolled in kindergarten, despite living in a household largely devoid of books. Early on as a child Rosalyn had been interested in obtaining a doctorate and had even said as much to an older relative. Rosalyn Sussman has been described by biographers as a child with a strong will, independent, stubborn, and determined.

The story is told that during Rosalyn's first day of elementary school, she hit her teacher back after being struck on her hand with a ruler. Young Rosalyn then explained in the principal's office that she was avenging her brother who had been similarly hit on the hand by the same teacher 5 years before. Apparently, her parents were described as being shocked yet proud and somewhat amused by the incident and their strong-willed daughter. Her father, in particular, was reported to have regularly encouraged Rosalyn's combative nature, a character trait that was to be instrumental in garnering later success during her scientific career.

In middle school, a teacher informed Clara Sussman that her daughter was a genius. Rosalyn graduated from high school in 1937 at the age of 15, two years ahead of time.

2) Apparently she earned a B.A. in chemistry and physics (magna cum laude!) from Hunter College, in New York City. This was quite an accomplishment at the time. What did she decide to do after graduation?

Rosalyn's academic performance in Walton High School, located in the Bronx, NY, qualified her for a full scholarship to attend Hunter College, a prestigious institution, and her attendance there would have been virtually impossible because of her precarious financial situation at the time. While in college Rosalyn Sussman had had an interest in becoming a physician, but at the time, being a Jewish female, it was all but impossible for her.

Rosalyn's other interests included physics and mathematics, especially calculus. As the institution at the time had no physics program as an academic major, the school's administrators, upon her request, had developed such an academic discipline, and Sussman was one of the first students to sign up and graduate from the fledgling physics program. She had felt fortunate in that she was able to personally attend as a female undergraduate student in 1939 a seminar given by the famous Prof. Enrico Fermi, Nobel Laurate in physics who was instrumental in splitting the atom and later in designing the atomic bomb.

A major influence upon the young Rosalyn Sussman was her reading in 1937 of the biography about Dr. Marie Curie, written by Curie's daughter Eve Curie. Rosalyn found the book to be inspirational as it demonstrated to her that perseverance would most certainly pay off in the end. In 1941at the age of 19, Rosalyn Sussman took her undergraduate bachelor degree of arts in physics and chemistry, with, as you pointed out, high academic honors.

3) During that time, the sciences were typically all-male. But what occurred to assist her in following her dreams — would she ever get a job in her field?

Gaining entry into graduate school proved to be an extremely difficult task for her, due primarily to anti-Semitism and widespread sexism. Purdue University responded to her graduate school application by stating they could not accept her unless she had already had in hand a job lined up afterwards. She did not, and therefore, Purdue did not offer Sussman a graduate assistantship.

Unfortunately, her undergraduate faculty advisors did not encourage continued higher education for graduate school. Instead, they steered her in the direction of secretary school to learn stenography and, thus, to become a secretary, all of which she did. She became a secretary at Columbia University in their college of Physicians and Surgeons, taking science courses part-time.

It was during this stint in 1941 that she was offered a graduate teaching assistantship from the physics department in the school of engineering at the University of Illinois, in Champaign-Urbana, Ill. Being female, she was not at first permitted to teach physics to the male engineering students, and she was, instead, allowed to teach only the so-called less qualified pre-medicine students. Student housing was for non-Jewish males only. Worse, there were no dedicated restrooms for women in the buildings where laboratory courses were taught. While in graduate school, Rosalyn saw the then popular black and white film *Madame Curie*, starring the well-known actress Greer Garson portraying Marie Curie and the well-known actor Walter Pidgeon as Pierre Curie. In 1942, Rosalyn took her M.S. degree in physics from the University of Illinois.

In Ph.D. school, Rosalyn worked in the laboratory of Dr. Maurice Goldhaber and met her future husband Aaron Yalow, also a graduate student in the Goldhaber lab. Interestingly, Dr. Gertrude Goldhaber, Maurice's wife, a notable and prominent physicist who was unable to hold a faculty position, due to university policy about nepotism—two relatives could not work at the same university—had become an ardent supporter and mentor to the younger Sussman.

Rosalyn and Aaron were not permitted to marry because they had been granted teaching assistantships and were then in effect also two faculty working at the university; they somehow managed to circumvent the requirement when Aaron received an outside fellowship—they were married on June 6, 1943. Working in the research laboratory late at night she sometimes slept on the floor of the bathroom!

Earning almost all perfect grades in her graduate-level courses, with the one exception of an "A-" in her laboratory course in electrodynamics, she was nonetheless called into the physics department chair's office and told that her less-than-stellar grade in the course showed that "women do not do well with respect to laboratory work." She chose not to argue with the chair as he was only an acting-chair.

Yalow nevertheless pursued her thesis project which involved building equipment for measuring the amounts of radioactive materials. For her thesis project, Yalow characterized how the element tellurium radioactively decayed to the element antimony, firmly establishing the mechanism of the conversion. Dr. Rosalyn Sussman Yalow successfully took her Ph.D. degree on the month of January in 1945 in the field of nuclear physics from the University of Illinois.

Obtaining a permanent academic position was also a difficult task. Thus, Dr. Yalow took a non-academic job in 1945, as the only female engineer at the Federal Telecommunications Research Laboratory housed at the IT&T Corporation. In 1946, Dr. Yalow became a physics lecturer at Hunter College, her alma mater. She was subsequently promoted to assistant professor of physics. Dr. Yalow also became a scientific consultant at the Bronx Veterans Administration Hospital and was key to establishing a facility devoted to performing radioisotope services for medical treatments.

It is here where Dr. Yalow met her colleague and collaborator Dr. Solomon Berson. While at the VA hospital as director of her active research laboratory, Dr. Yalow also became distinguished professor-at-large at other institutions, such as the Albert Einstein College of Medicine and the Mount Sinai Medical Center, CUNY, in New York.

4) As far as I know Yalow is most known for the development of a specific technique for analyzing human blood. I know some of the basics - Red blood count, white blood count, total protein, hematocrit, HDL (healthy cholesterol) LDL (Lousy Cholesterol) but who was the first to really examine blood- and how is it done - and what did Yalow specifically do to further blood analysis?

The study of blood goes back two millennia, in ancient times, with the notion of the so-called four humors—blood, phlegm, yellow bile and black bile. The humoral theory held that disease was caused by, for instance, imbalances in, or the fermentation of the various four humors. Of these, the early theories of blood were shown to be the most accurate, at least from a medical standpoint.

Probably the first physiological studies of blood had occurred in 1628 with Dr. William Harvey, who studied the blood flow circulation system. In 1837, Gustav Magnus was the first to detect blood gases, such as oxygen, carbon dioxide, and nitrogen. In 1900, Dr. Karl Landsteiner discovered the A, B, and O blood group antigens for determination of blood types. In the 1920s, the hormone insulin was discovered to be associated with diabetes, a condition noted by abnormal levels of sugar in the blood as regulated by insulin.

Working together, Drs. Yalow and Berson developed the so-called RIA (radioimmunoassay) technique to study how long insulin remained in the blood of diabetic humans. They labeled insulin with a radioactive isotope, the radioactive energy of which could be measured readily by Yalow, then injected patients with the radiolabeled insulin, taking constant blood samples and finding out how long it took for the radioactive hormone to leave the body's blood circulatory system. They found that in diabetic patients, insulin left their systems more slowly than in non-diabetic patients. The explanation for this observation was unclear at the time—and Yalow and Berson were the first to provide the correct explanation.

They reasoned that the injected insulin was generating an immune response and that possibly the human body was making antibodies against the foreign insulin. This interpretation was met with disbelief because other investigators believed that insulin was simply too small of a molecule to be antigenic. It turns out, of course, that Yalow and Berson were correct: insulin was antigenic and an immune response could be mounted against the hormone. The antibodies somehow neutralized the effects of the injected insulin, making it difficult to process the blood insulin, and this is why it took longer to leave the circulation.

This work was nonetheless summarily rejected by the prestigious journal *Science*, and the second choice, the *Journal of Clinical Investigation* would publish the result, but only if they left out the notion of an "anti-insulin antibody" from the manuscript—Yalow and Berson dutifully took out the offending terminology, and the paper was published, leaving Yalow to be quite annoyed about it for many years to come—because they had been correct all along about insulin being both immunogenic (generating an immune response) and antigenic (reacting with immune cells or antibody—in this case, an antibody).

With the hormone insulin established as being a good antigen, it led Yalow and Berson to follow-up on the discovery to develop their Nobel Prize winning invention: the RIA.

First, they labeled the antigen insulin with a radioisotope, iodine-131 or iodine-125, by synthesizing the antigen in the presence of radiolabeled amino acids. Next, they mixed the labeled insulin antigen with antiserum, containing anti-insulin antibodies, to form an antigen-antibody complex.

Then, they added Protein A, which binds to the antigen-antibody complex, forming a solid precipitate. Next, they centrifuged the material forming a supernatant on the top of the test tube and a "pellet" on the bottom of the test tube. With the labeled antigen-antibody-Protein A complex being relatively heavier than the antigen alone, a positive test meant that the nature of the antigen and antibody were complementary and the pellet would contain the bulk of the radioactive iodine-131. A so-called negative test would detect the unbound antigen in the supernatant.

The next experiment clinched the importance of their RIA invention. Yalow and Berson then asked how much unlabeled antigen they could add to disrupt the labeled-antigen from the antigen-antibody complex. This amount would be indicative of the total amount of antigen that was detectable.

To their astonishment, Yalow and Berson found that their RIA technique was extremely sensitive and could detect vanishingly tiny amounts of material—about a half to one nano-gram, i.e., on the order of a billionth of a gram! This achievement has been described as the equivalent of being able to detect a teaspoon of sugar in a lake of water that's 9 meters deep, and 100 kilometers long by 100 km wide! That is to say, the RIA technique could detect extremely minute quantities of just about any physiologically relevant agent.

5) Apparently her work had a lot to do with modern physiology- In what way were her discoveries relevant and salient?

I think one of the first immediately important outcomes based on the discovery of the RIA technique was now the capacity for biomedical researchers and clinicians alike to measure the blood levels of the insulin hormone. This was an important advancement in learning about the physiology of diabetes. Soon, other physiologically important molecules were detected in minute quantities in patients.

Such biological molecules that were detected by the RIA technique early one included serum albumin in patients with congestive heart failure; human growth hormones in patients with deficiencies in the hormone; adrenocorticotropic hormone (ACTH) levels in patients with ACTH deficiency or in patients with too much ACTH, as seen in cases of lung tumors or in Cushing's disease, a condition involving pituitary gland tumors.

Early on Yalow and Berson studied other biologically important molecules that were amenable to evaluation by the RIA method and included parathyroid hormone levels in hyperparathyroidism patients; and they studied gastrin hormone concentrations in patients with peptic ulcers. Each of these involved important advances in their fields.

Soon after the publication of these early studies by Yalow and Berson, the list of suitable molecules that could be analyzed by RIA grew exponentially, from other protein-based hormones to non-protein (small molecule) hormones, to intermediate metabolites, to many other molecules.

Additionally, the RIA invention could be used to measure the immune response in higher organisms by detecting and quantifying the concentrations of antibody produced or the production of immune-specific molecules such as cytokines and mediators, that result during an immune response to antigens, such as microbes like bacteria or viruses, and cancer, or graft transplantations, like those of cells, tissue and organs. Thus, the RIA technique is a key and enduring staple in the field of immunology.

In short, of the many tens of thousands of distinctive molecules that reside in the body and which are needed for the proper functioning in order to maintain life, each of these molecules must be present in the correct locations within the body and in their proper amounts. Thus, the list of agents that could be detected and quantified by RIA was and is virtually endless. As a result, the RIA technique of Yalow and Berson continues to be included in modern textbooks pertaining to biochemistry, physiology, immunology and endocrinology, just to name a few.

6) In addition to the Nobel Prize in Medicine- what were some of her other accomplishments?

Because of the universal application of the RIA invention to all branches of the medical, biomedical and basic biological sciences, Dr. Yalow's accomplishments in these other areas of science were far reaching. Within the field of diabetes research, she discovered that insulin levels in the blood plasma were similar if not higher in diabetics, thus showing that diabetes was due to the inability to utilize insulin, rather than as a result of depleted insulin levels.

Dr. Yalow was also instrumental in the study of vitamin B12 metabolism in humans. Additionally, because of her work in detecting hepatitis virus in the blood using the RIA, blood donation banks were able to adopt the procedure of routinely evaluating donated blood for the presence of the hepatitis virus microbe in their blood samples. Her laboratory had demonstrated that the gastrointestinal hormone, cholecystokinin, not only played a key role in the metabolic degradation of fats in the gastrointestinal tract, it was also functional in the brain as an important synaptic neurotransmitter, thus, linking the physiology of the GI tract to the brain.

In my mind, another important achievement constitutes Dr. Yalow's amazing abilities to repeatedly overcome entrenched sexism and anti-Semitism during her lifetime in order to become scientifically successful. On the other hand, most, if not all, of her male colleagues did not have to expend so much energy and efforts just to be able to conduct their scientific activities.

Dr. Yalow was widely celebrated, with entry into the prestigious National Academy of Sciences and recipient of the Albert Lasker basic medical research award, the latter of which is considered a preamble to the Nobel.

7) How were the last years of her life spent?

Although Dr. Yalow spent her later years as a retired investigator, she nevertheless continued to be active in scientific circles. One issue that was important to her was about the laboratory regulations that were in place for the use of radioactive materials. She felt there was a significant difference between small doses of low-energy radiation as conducted in biomedical science laboratories and chronic doses of high-energy radiation in nuclear power and energy research, yet, the procedures for both types of work were largely the same.

One incident conveyed by Yalow highlighted how radiation overregulation had become widespread. A laboratory radiation inspector had criticized her RIA procedure manual as inadequate—to which she replied merely "I wrote the manual." Indeed, she had invented the RIA.

During her later years, Dr. Yalow also took to the lecture circuit and became an activist for various scientific causes. She spoke of the importance of improving American education, in establishing child-care services for women trying to get an education in colleges and universities, and issues pertaining to women in science. She spoke of the disparities in salaries between male and female scientists in which women earn significantly less than their male counterparts for similar work. Dr. Yalow expressed the important need for equal opportunity with respect to women scientific investigators.

8) I must say that as a woman in the 1940's and a Jew, she certainly had her share of troubles overcoming that cultural milieu. Any insights into her personality?

Biographers have noted that the attributes of her personality that she had acquired in childhood, such as determination, strength of character, steadfastness, and so on, were to remain with her in adulthood, as well. It has been speculated that these and other attributes provided the skills sets required to overcome the obstacles inherent in established and deeply ingrained anti-Semitism and sexism that she had no doubt encountered.

The situation was so prejudicial that, for instance, when Berson experienced an untimely and tragic death, Yalow had to prove herself all over again, as many critics at the time had incorrectly attributed their successes primarily to Berson. She went on, however, to publish numerous papers in peer reviewed scientific journals for many years afterwards, clearly establishing herself as a dedicated and stellar independent research scientist in her own right.

Furthermore, I think it's unfortunate that individuals with whom Yalow may have disagreed with became her harshest of critics, even becoming personal in their attacks of her. One may surmise whether such harsh criticism would have been forthcoming had she been a male scientist. Indeed, such vitriol is relatively rare in cases where the disagreeing individuals are male. To those with whom Yalow was close to and who knew her well, she was known to have regularly conducted herself with grace, kindness, and humility. I think that Dr. Yalow should be commended not only for her scientific achievements but also for those accomplishments in overcoming adversity.

9) What have I neglected to ask about this famous woman scientist?

You'll recall I mentioned above that Dr. Yalow's early work regarding insulin being antigenic and thus being able to bind an antibody specific for it, had been met with widespread disbelief among the experts in the field. Of course, Dr. Yalow had been totally correct about the ability of an anti-insulin antibody to bind insulin. The antibody-insulin association has been convincingly confirmed since then with many scientific papers providing clear and supporting evidence attesting to the validity of the idea.

The story is told that soon after receiving the Nobel, Yalow published her ceremonial Nobel Prize lecture from Stockholm in the journal *Science* and included an excerpt of the letter from the editor of the *Journal of Clinical Investigation* advising her to remove the antibody-insulin connection in the manuscript. She was at the time, 1978, widely criticized for having done this.

Unfortunately, in 1972 Dr. Yalow's colleague, Dr. Berson, had died from a massive myocardial infarction while attending a scientific conference. As a consequence, he was not able to share in the receipt of the Nobel with Yalow in 1977. According to the Nobel Prize committee and policies, the prizes are bestowed only to living individuals. As a tribute, Yalow made arrangements to have her research laboratory named after Berson.

I mentioned earlier that there had been widespread concern about the use of radiation in common laboratory methods, such as the RIA. Because of the fear of even low levels of radiation use, non-radioactive techniques were developed, but had been based on the same principles as those used for the RIA.

One of these relatively newer methods uses enzymes that are linked to antibodies in which a positive result shows a colorless but chromogenic enzyme substrate turning into a colorful product upon successful antibody binding to an antigen. One method in use today is called the ELISA (pronounced "Eliza"), which stands for Enzyme-Linked Immunosorbent Assay. The procedure, commonly used today in many branches of biological, physiological, biochemical, immunological and biomedical sciences, is a non-radioactive method for the sensitive detection of molecules, and it is, thus, regarded as safer to use than the RIA.

Chapter 34

Har Gobind Khorana

1) Professor Varela, we have all heard about the movie The Da Vinci Code, but probably an even more important code was the "genetic code." So, let's start at the beginning. When we discuss a "genetic code" what are we really talking about?

Like Tom Hank's fictional character, Dr. Robert Langdon, in the book and film *The Da Vinci Code*, Prof. Har Gobind Khorana, was also involved in momentous efforts aimed at solving an extremely complicated genetic code puzzle, with epic and profound implications in its solution. The genetic code has essentially been life's greatest molecular secret.

The genetic code can be thought of as a set of biological blueprints for specifying the production and workings of all living beings on the planet. The genetic code is embedded within the extremely minute and seemingly unending DNA chains of nucleotide sequences that constitute everyone's genome, including all living things and all life that has ever existed in Earth's history. Simply put, without the genetic code there would most certainly be no life.

The genetic code that's part of the DNA molecules is converted by transcription to make an RNA transcript code that's then used, in turn, for the production of proteins in a process called translation and which connects various lengths of amino acid chains to compose these proteins. The protein chains then fold into functional three-dimensional shapes that then perform biochemical reactions or form support structures for the inner workings of cells that make up organisms.

The solving of the genetic code constitutes one of the greatest achievements in the fields of Biology, Molecular Biology, Genetics, Genomics, Proteomics, Biochemistry, Bioinformatics, and many other disciplines. In the year 1968, the scientific researchers Drs. Marshall Nirenberg, Robert Holley and Har Gobind Khorana travelled to Stockholm to receive the Nobel Prize within the fields of Medicine or Physiology for their experimental efforts conducted towards finding out the secret to the genetic code that's present in DNA molecules.

2) Now who was Har Gobind Khorana, and where and when was he born?

The actual date of Har Gobind Khorana's birthday is unknown; and, in documents reported by Khorana in later years, they show an estimated date of the 9th of January in the year 1922, although it is unclear how this date came about. Khorana was born in the village of Raipur, British India, in what later became Western Punjab province and is now Raipur, Western Pakistan. Khorana's father had been a government employee, in the governing system that had been set up by the British Indian government, working tirelessly as an agricultural taxation clerk ("patwari") for the tiny village with only about 100 residents living in it. Khorana's Hindu parents, Shri Ganput Rai Khorana and Shrimat Hrishna Devi Khorana, were poor, and young Har Gobind along with his four older siblings grew up in poverty. Nevertheless, Khorana's parents had firmly instilled the value of an education early on in their progeny such that the Khorana children were virtually the only ones in the village who were able read.

3) How did he fare in school as a youth and what were his early educational experiences like?

Khorana attended high school at the D.A.V. (Dayanand Anglo-Vedic) high school located in Multan, present day West Punjab. In later years, Khorana reported that one of his teachers, Ratan Lal, at the D.A.V. school played a major influential role in the young Khorana.

Khorana later received a scholarship to study chemistry while attending Punjab University, also known as Government College University, located in Lahore, in the country of India. In 1943, Khorana received his B.S. undergraduate degree in chemistry, and in 1945 he took his master's degree in science (M.Sc.) at the same institution. Khorana was to write in later years that another influential teacher at the school was Mahan Singh. I think it is quite telling about Khorana's character as a grateful person in that he often credited his teachers and mentors. Khorana had earned both university degrees with full academic honors, thus, earning a graduate fellowship from the government of India to attend graduate school in England at the University of Liverpool, where he entered the laboratory of Prof. Roger J.S. Beer. It had been the first time that Khorana had left his native country. In 1948, Khorana successfully defended his thesis and received his Ph.D. studying the organic chemistry of melanin compounds.

The newly minted Dr. Khorana then became a postdoctoral student studying the chemistry of alkaloid chemicals in the laboratory of Dr. Vladimir Prelog, future Nobelist (in 1975), at the Swiss Federal Institute of Technology (ETH) in Zurich, Switzerland.

Khorana's tenure in Prelog's laboratory at the ETH was less than a full year, as there was little to no funding for Khorana; he apparently used his bank savings, subsisted primarily on unpasteurized milk and rice, and lived in the laboratory. Nevertheless, Khorana credits Dr. Prelog with having trained him (Khorana) well in terms of work ethic and various approaches and strategies for conducting scientific research. Interestingly, Khorana's work in the laboratory was not productive, driving him to read up on the literature where he came across an obscure publication about a reagent called carbodiimide, which he would later use to make a history-changing Nobel Prize-worthy discovery in later years.

Dr. Khorana spent the fall of 1949 in India paying service back to the government for having conferred the necessary funding to attain his education abroad; but because of political changes that had occurred in the meantime and the altered the Indian-Pakistani borders, it was difficult for Khorana to find suitable employment. In 1950 Khorana returned to England where he received a postdoctoral fellowship to conduct research at Cambridge University in the laboratory of Sir Alexander R. Todd, working with G.W. Kenner. Using the carbodiimide reagent, Khorana and Kenner activated the so-called carboxyl terminals (C-termini) at the ends of short peptides, which are chains of amino acids.

4) What led him into genetics (did he study Gregor Mendel)?

It is while working as a postdoctoral fellow in the Todd laboratory at Cambridge during the interval between 1950 and 1952 that Dr. Khorana became interested in the study of nucleic acids research. While it is uncertain whether Khorana knew of Mendelian genetics while a student at university, as he was educated as a chemist and not necessarily as a biologist, it is presumably while in the Todd research laboratory working with nucleic acid chemistry that Khorana must have somehow picked up the knowledge discipline pertaining to the field of genetics. There is no doubt, however, that Khorana was later to put a serious dent in the cracking of the knowledge base locked away in the mystery code that Mendel discovered.

Incidentally, it is during this time period that Khorana got married to Esther Elizabeth Silber, who was from Switzerland and whom he had met in 1947 while briefly staying in Prague.

It is also during his stint in Prof. Todd's laboratory that Khorana learned of an opportunity in Vancouver, British Columbia, Canada, where he would have no start-up funds for an independent laboratory, but he would, nonetheless, have freedom to study whatever he wanted in the laboratory of Prof. Gordon Shrum. In 1952, the Khorana family moved to the University of British Columbia. Khorana studied the chemistry of phosphates and nucleic acids.

Khorana then moved to a new academic position in 1960 at the University of Wisconsin, within the associated Institute for Enzyme Research, and in 1966 Khorana had become a citizen of the U.S. It is here where Khorana was noted to have conducted his work pertaining to the genetic code.

5) What exactly were his discoveries?

Khorana's participation in solving the genetic code inherent within DNA involved his expertise in synthesizing nucleic acid strings along DNA within the laboratory. Interestingly, it is here where Khorana's serendipitous knowledge of the carbodiimide reagent turned out to be advantageous in his DNA synthesis experiments. First, employing the use of the carbodiimide chemical reagent in the laboratory, he had figured out how to make ADP and ATP, both molecules of which contain one of the nitrogenous bases, namely, adenosine, found within the genetically important DNA molecule.

Khorana then used his chemical synthesis technique to produce so-called polynucleotides, later called oligonucleotides, which were long chains of nucleotides along a DNA molecule. This finding was to be of great importance in Khorana's work to help solve the genetic code.

In 1961 Khorana learned about a seminal experiment conducted by Dr. Marshall Nirenberg and his postdoc Heinrich Matthaei at the National Institutes of Health. In their paper they described how they added a so-called poly-U RNA molecule, containing a long chain of only uridine molecules in it, to a solution containing the insides of exploded bacteria, a solution known as a cell-free lysate or extract, containing all of the biological machinery needed to make proteins. The mixture of cell-free extract and poly-U RNA were added to separately purified amino acids, and the experiment produced a long chain of amino acids connected to each other consisting of only phenylalanine (called poly-Phe).

Thus, the triplet codon, UUU, encodes phenylalanine! Next, using poly-A, the investigators found that the AAA codon now specified the amino acid called lysine, and with poly-C the CCC codon stipulated the amino acid called proline.

These results intrigued and energized Khorana. With a stroke of genius, Khorana realized that his knowledge of making DNA could be used to solve the rest of the genetic code! He and his laboratory group used his efficient DNA synthesizing expertise at the University of Wisconsin in Madison to make DNA polynucleotide molecules of various sequences that were then used to make their corresponding polyribonucleotide RNA molecules of precise sequences. Khorana had to add DNA polymerase to make his DNA and RNA polymerase to make his RNA molecules. He also had to somehow figure out the sequences of the resulting protein chains produced by his RNA experiments.

The experimental approach used by Khorana could be used to produce repeating units of various sets of RNA codons. Khorana's work was laborious, working day and night. The laboratory produced RNA chains that contained all other possible triplet codons, e.g., AUC, UCA, CAU, UCU, CUC, etc., The list of the possible RNA codon iterations was exhaustive. Next, Khorana added each of the purified amino acids and examined which ones were incorporated into the protein chains being produced in the cell-free extracts. He analyzed all of the results of his repeating RNA codon sets for the various types of polypeptides produced. For instance, he discovered that the RNA codon AUC encoded the amino acid isoleucine. Likewise, UCA specified serine, and CAU encoded histidine.

He then proceeded to systematically examine all of the possible triplet codon combinations. In so doing, Khorana's work resulted in the complete unraveling of the rest of the genetic code!

He even found termination codes, which specified so-called stop codons, telling the protein synthesis machinery when to stop translation and make the end of a protein chain.

Khorana's work also contributed to the notion that the genetic code was degenerate. That is to say, more than one codon may specify an amino acid. The same is true of the stop codons, as there are three such codons that encode a protein chain termination.

6) Did he receive any acknowledgement for his discoveries?

Prof. Khorana was widely and soundly acknowledged for his revolutionary scientific discovery towards solving the mystery of the genetic code. Together, in 1968, the three principal investigators from each of the laboratories run by Drs. Nirenberg, Khorana, and Holley learned that they were to share the Nobel Prize in the important topic pertaining to Medicine or Physiology. Each of these investigators provided an important contribution to the solving of the genetic code mystery. Nirenberg started the puzzle solution to the genetic code; Khorana figured out the remaining triplet codons and their respective amino acids they encoded. He also demonstrated for the first time that there was an important connection between the DNA code and protein synthesis, with RNA directly serving as a genetically involved intermediate between them. Holley was the first investigator to purify an intact tRNA molecule, demonstrating that it was necessary for translation, and to elucidate its sequence and structure.

7) Why is this genetic code relevant to scientists or is it just relevant to geneticists?

It is my firm contention that the genetic code is not only relevant to geneticists and many other scientists alike, but that it is also relevant to just about everyone. My reasons for this are as follows. First, on a fundamental level, knowledge of the genetic code represents knowing something about the molecular secret to life. As it happens, the genetic code that Khorana and others helped to elucidate is a largely universal code; that is, with the exception of the viruses, just about all other living beings, from bacteria to humans, share the same genetic code.

After the solving of the genetic code was performed, it became possible to understand how gene activities and expression are controlled. Such gene expression regulation allowed investigators to artificially turn on genes when needed or to turn off genes when necessary.

The genetic code has helped us learn about where we came from as a life force and perhaps has provided us with some clues about where humans are headed to in the future. With this knowledge in hand, one is able to use it to address the problem of hereditary diseases in order to find ways to detect, diagnose, treat, and maybe even correct a malfunctioning or defective gene. The genetic code may be used to produce needed medicines or important products that improve the quality of life.

Today, the genetic code elucidation has allowed investigators to determine the code sequence of nucleotides along the DNA genome chains for individuals. These genome projects are numerous and provide a tremendous amount of information about an individual's phylogenetic evolutionary history, physiology, biochemistry, cellular biology, molecular biology, immunity, resistance to harmful substances, and even pathological pathways.

8) What were his later years spent doing?

In 1971, Khorana moved to the prestigious Massachusetts Institute of Technology (MIT), in Cambridge, MA. He eventually was promoted to the endowed Alfred Sloan Professor of Biology and Chemistry at MIT, where he continued to work until his retirement in 2007.

Some of his other work involved the making from scratch of the world's first artificially produced gene, which encoded a yeast tRNA gene specific for the amino acid alanine. It was a massive undertaking, involving many laboratory co-workers headed by Khorana, and published in a single issue of the *Journal of Molecular Biology* as 15 research papers in December of 1972.

An even more impressive achievement from Khorana's laboratory was the synthesis of an artificially produced DNA molecule encoding a suppressor tRNA gene that actually functioned as a proper gene when placed inside a new host bacterial cell, in the late 1970s.

His later years were spent pursuing interests in cellular signaling within the eye involving a molecule called rhodopsin, a light sensitive protein molecule embedded within the biological membrane, and associated mutations of which are implicated in a night blindness causing condition known as retinitis pigmentosa, a genetic disease found in humans. Interestingly the rhodopsin molecule is evolutionarily conserved in all domains of life, from bacteria to humans. This was some of the last research work that had interested Khorana in his later years. At the age of 89, Khorana died on the 9th of November in the year 2011.

9) What have I neglected to ask about this fascinating scientist?

Apparently, Prof. Khorana was a shy and unassuming person who tended to avoid the limelight and never wanted any fuss made on his behalf. He especially did not like using a phone. An interesting anecdote was conveyed by one of Khorana's colleagues, Dr. Thomas Sakmar, which was common knowledge among his close friends and colleagues but not known to others outside of his immediate circle.

In 1987, he was to receive the prestigious National Medal of Science, but because he avoided the phone, the important phone-messages informing him of the honor piled up on his desk, and he was not aware of the impending ceremony and award, soon to be bestowed at the White House in Washington, D.C. As a result, a White House official had to travel to a scientific conference where Dr. Khorana was attending and track him down to inform him of the logistics for receipt of his award at the White House.

Another anecdote involves myself and the one time I ever laid eyes on Prof. Khorana, himself. It was while I was a postdoctoral fellow at Harvard Medical School in the mid-1990s. Once in a while, I would take the Green and Red Lines, Boston MBTA (Massachusetts Bay Transportation Authority) routes from Longwood Avenue where the medical school was located to MIT at Kendall Square in Cambridge to visit with my good friend and MIT biochemistry graduate student, Angelo Mondragon, for lunch, to talk science, and to visit the MIT bookstore. Being that Harvard pizza was not very good, I visited Angelo as often as I possibly could as he had assured me that MIT pizza was better—it was. It was during one of these MIT cafeteria visits that Angelo pointed out the famous Dr. Khorana, who was standing in the cafeteria line purchasing MIT pizza.

Chapter 35

CARL WOESE

1) Dr. Varela, I have always believed that scientists often think differently than other normal folk. And apparently, Carl Woese also believed that the minds of scientists were somewhat more organized. So, today, let's look at the factors that make up the mind of a scientist. First of all, what do we know about where he was born, and when and his parent's environment.

Prof. Carl Richard Woese, an American biophysicist and bacteriological microbiologist, was born on the 15th of July in the year 1928 in the city of Syracuse, New York. Woese's middle-class parents were Carl Frank, an engineer, and Gertrude Nadler Woese.

2) His early education- is there any significant record or things that stand out?

As a high school student in the 1940s, Woese was enrolled in the prestigious Deerfield Academy, a boarding school founded by Samuel Adams in 1797 and housed in the town of Deerfield, MA, in the U.S. The school was a largely college preparatory institution. Many of its graduates went onto Ivy League colleges and universities. In a popular book, Woese mentions that in K-12 schools in the U.S., students are taught about Petri dishes and culturing bacteria, or how to grow plants, or about the importance of environment. Presumably, Woese may have experienced these and other concepts as a student.

3) Where did he go to college? I know that some colleges teach higher order thinking and critical thinking more extensively than others.

After high school graduation, Woese attended Amherst College, a small and privately-owned liberal arts institution, situated in the city of Amherst, MA. Majoring in physics and mathematics, Woese took his Bachelor of Science degree from Amherst in 1950. While an undergraduate, his academic advisor encouraged Woese to pursue the study of biophysics. Entering the research laboratory of Dr. Ernest Pollard at Yale University, Woese became a virologist and studied the physical effects of heat and ionizing radiation on hemagglutinin from animal viruses, such as the rubella virus, avula virus, and influenza virus. Woese earned his Ph.D. in the field of microbiology from Yale in 1953, and published these works starting in 1954 at the age of 26 in the journal called *Archives of Biochemistry and Biophysics.*

As a postdoctoral fellow at Yale, Dr. Woese studied the induction of certain viruses, called bacteriophages or phages, which infect the endospore-producing bacterium called *Bacillus megatarium* during the process of germination when a bacterium is converting from an endospore to an actively growing and metabolizing state. The phage induction involves converting the phage from an integrated state, in which the phage genome is inserted into the bacterial genome, to a lytic state, in which the virus multiplies inside the germinated bacterium, producing many viral progeny and lysing the bacterial cell in turn. Dr. Woese published this work in 1960.

After briefly working at General Electric Research Laboratory as a biophysicist, Dr. Woese became a faculty member at the University of Illinois located in Urbana-Champaign, IL, later called the University of Illinois Institute for Genomic Biology, where he remained for the rest of his academic career.

4) What is this "three domain theory of taxonomy"? And how does it apply to science?

Essentially, in 1977 Carl Woese proposed the presence of a brand new domain called the Archaea and that all living organisms on Earth could be grouped into three fundamental domains: the Archaea, the Bacteria and the Eukaryotes. The Archaea consist of so-called ancient bacteria, most of which can be found in extreme environments, like extreme heat or cold, or in acids, salts, etc. The Bacteria domain consists of the rest of the prokaryotes. The Eukaryotes domain consists of every eukaryotic life form: all of the animals (including humans), plants, fungi, and protists.

The Three Domain system of classification (or taxonomy—the science of classification for living organisms first developed by Carolus Linnaeus) came about starting with Darwin's evolutionary biology, the notion of a universal common ancestor, and subsequent groupings of related organisms into taxa (groups). These groupings into taxa allowed a naming system for organisms so that investigators anywhere would know and understand what one means when, e.g., *Escherichia coli* is stated in the literature or at a scientific conference; everyone is brought to the same page when an organism's taxonomic name is mentioned. Secondly, the Domain system facilitates learning new pieces of biological information about newly discovered organisms. The term phylogeny refers to evolutionary history. Invoking the knowledge gained from genetics, the term phylogenetics has emerged to describe the inherent evolutionary relationships.

In the 1950s and 1960s, the fields of molecular-based evolution and bioinformatics were developed. These branches started with Frederick Sanger, who developed techniques for determining the sequences of amino acids along a protein chain and later for determining the sequences of nucleotide sequences along a DNA chain. Prof. Sanger earned two Nobel Prizes for Chemistry, one in 1958 for his studies of insulin structure using its amino acid sequence, and the other in 1980 for developing his technique for determining DNA nucleotide sequences.

Then, during the 1960s, Linus Pauling and his collaborator Emile Zuckerkandl theorized that these protein and DNA sequences could be used to study evolution. The idea was that closely related groups of organisms would have related protein and DNA sequences. This is where Carl Woese enters the picture. Woese began to study the DNA sequences of genes that he felt all living organisms should harbor, namely, DNA genes encoding ribosomal RNA, found in ribosomes. All living organisms have ribosomes, for making proteins.

Dr. Woese collected DNA samples from all of the living beings that he could get his hands on and sequenced their rRNA gene sequences. The work was painstaking, slow, and labor-intensive. What he found after all of his hard work, however, greatly surprised him!

First, he found that the Archaea and the Bacteria, both of which are prokaryotes, were, nevertheless, distantly related to each other! Then, even more shockingly, he found that the Archaea were more closely related to the Eukaryotes than to the Bacteria! On a molecular level all life could be grouped into the famous Three Domains.

Confirmation of the Three Domain idea was established with the sequencing of the entire genome of an Archaea microbe in the 1990s. The sequence comparison results that Woese had found with rRNA genes from many different organisms all held up to scrutiny when entire genome sequences were determined in organisms from all three domains and compared to each other.

The ramifications and consequent importance of Woese's Three Domain concept were profound. First, the molecular sequence and phylogenetic evolutionary data were used to construct a so-called tree of life, encompassing all known genomic sequence data. The tree of life pointed to a common ancestor for all life on Earth, described as the Last Universal Common Ancestor (LUCA). The LUCA concept holds up whether the phylogenetic trees are built using genes or genomes.

The tree of life also shows how modern eukaryotes have mitochondria that are derived from bacteria! All eukaryotes have mitochondria, which are sub-cellular organelles that make ATP energy for life's processes to occur. For instance, the human mitochondrial DNA is unlike any other eukaryotic organism's DNA; it's more bacterial than human!

It is widely thought that 3.8 billion years ago, during the primordial conditions of Earth, the Archaea emerged and invented new biological mechanisms that later life forms absconded with, such as enzymes that neutralize the dangerously oxidative oxygen and its equally harsh derivatives, or invented photosynthesis, or developed cellular mechanisms for the synthesis of molecular building blocks of life, such as nucleotides or amino acids. Because of these and other inventions made billions of years ago by the microbes within the Archaea domain, life on Earth became possible for eukaryotes, too, such as humans.

Further, if DNA is collected from an individual, e.g., a bacterium or a person, or if DNA is taken from an environment, e.g., soil or sea water, etc., that harvested DNA, with all of the individual genomes from the organisms present in the samples, can now be easily sequenced, then compared to the known sequences already stored in computer databases and all of the organisms identified to the species level. This new field is called meta-genomics. Because of Woese's work, to this day whenever an investigator wishes to identify a newly discovered micro-organism, he or she must look at the rRNA gene sequences in order to do so.

This is especially important in microbial ecology, where only an extremely small fraction of the known microbes have been discovered, thus far. Recently, it was proposed that there are at least a trillion microbial species! Clearly, therefore, there is much more work to be accomplished if we want to know and understand who all of the microbes on Earth are. Some of these yet undiscovered microbial species may prove to be extremely beneficial.

Another ramification of Woese's work was that using the protein sequences of amino acids, the three-dimensional structures of the proteins could be predicted and studied. This will permit investigators to learn the biochemistry of these molecules and to use them for improvement of biomedical and biotechnological products, such as in diagnostics, advances in new medicines, or in understanding how normal physiology and biochemistry happens versus how disease states manifest themselves.

Furthermore, using computers to analyze DNA or protein sequences, a field known as bioinformatics was formed. The computer analysis of protein and RNA sequences could be used to identify key evolutionarily conserved structures within these molecules that are important for determining their important functions. If evolution has kept a gene, protein, RNA, or their structures throughout time, then bioinformatics can be used to identify these so we may learn new and important cellular and molecular mechanisms.

Thus, if a new gene is discovered, and its function is unknown, a bioinformatics work-up could help identify its function. Additionally, in molecules that are homologous, i.e., they share a common evolutionary origin, their physiology could be studied closely and used to develop new therapies for disease treatments, etc.

5) What else was Carl Woese known for?

Another important field in the biological sciences that Woese had a hand in involves the study of the origin of life on Earth. The interest is in learning how life on Earth began.

Towards this, Woese, along with biologists Leslie Orgel and Francis Crick, proposed a so-called RNA World theory, in which RNA arose first in the ancient primordial Earth, and that the ancient RNA could form three-dimensional structures and act like enzyme catalysts to make new molecules, like other RNA molecules and ribosomes, etc. From the RNA world, the DNA then took over as it is more stable.

Woese is also known for the concept of the so-called "progenotes" which are pre-cells with poor protein making machinery and lots of mutations in their DNA, and which may facilitate movement of new DNA between them and other cells, a process referred to as horizontal gene transfer.

6) What accolades did he receive when he was alive, and where was most of his work done?

Prof. Woese received many accolades during his life, such as entry into the National Academy of Sciences in 1998, receipt of the National Medal of Science in 2000, and the prestigious Crafoord Prize, in 2003. It is unclear, however, why Woese did not become a Nobel Laureate, as well.

The bulk of Woese's work was conducted while he was a faculty member at the University of Illinois and the Institute for Genomic Biology. After Woese passed away on the 30th of December, in the year 2012, the institute was named after him, the Carl R. Woese Institute for Genomic Biology.

7) His later years- what did he spend his time doing?

Woese worked all of his life and in later years he became philosophical, writing books and review articles. He speculated about the evolution of non-living entities and in identifying new and future areas that he felt needed further study, such as the ecology of the planet, the usefulness of microbial ecology, finding solutions needed to solve environmental problems, and the role of the biological sciences in society.

8) What have I neglected to ask about this fascinating scientist and the way his mind seemed to work?

It may be surprising for some to know that Carl Woese, a molecular evolutionary biologist whose work has permanently altered the way authors write the textbooks, was not an atheist. He was a firm believer in God, but was leery of organized religion and of people claiming to be religious. Furthermore, he was greatly suspicious of anyone using evolution in order to argue for or against the existence of God. Woese felt that the faithful understood God as a mystery, and the truly faithful required no proof of God's existence; and Woese understood that science was based on evidence—thus, the two areas are essentially unmixable, and should thus be kept separate. It is fascinating that many organized religions have no problem with evolution, especially if one knows that evolution involves simply the changing of DNA from generation to generation.

Chapter 36

SIDNEY ALTMAN

1) Professor Varela, first of all, while people may have HEARD about DNA—what exactly IS it, and why is it important?

Thank you for inquiring about one of my favorite molecules. In short, DNA is considered by many to be the principal molecule of life. As such, DNA can be envisaged as a never-ending living molecular blueprint for specifying all life on Earth.

DNA is a storage form containing a set of biological directions, so to speak, to confer life upon molecules. This famous DNA molecule directs certain biological functions within all living cells. Along these lines, DNA has associated with it certain molecular biological machinery for mediating the transcription process, also known as RNA synthesis. That is, DNA participates in its conversion of the genetic information stored in it to a messenger RNA transcript (mRNA) version of the code. The mRNA transcript version, in turn, works in conjunction with other forms of RNA, like ribosomal RNA (rRNA) and transfer RNA (tRNA) to perform translation for the purpose of producing gene products, such as proteins.

Hence, the expression of a gene involves the production of RNA and then protein, encoded within DNA. These proteinaceous gene products then perform their biological functions to produce and maintain new progeny cells and tissues contained within intact living organisms.

In addition to specifying molecular blueprints for living organisms, DNA also has molecular mechanisms for propagating its genetic blueprint information into the next generation, and the process occurs in the cells of all living beings on Earth. The genetic information propagation process also involves the synthesis of new DNA, also called DNA replication.

Many scientific investigators have used DNA to determine, compile, and annotate its nucleotide base sequences, placing the acquired genetic information into very large computer-based databases, for the purposes of identifying gene functions and of facilitating the study of protein biochemistry, physiology, and cell biology. DNA is a central molecule in the biomedical sciences discipline.

DNA has additional relevance. For instance, scientific investigators have used DNA as a tool for the basic and applied study of life. In these regards, molecular biologists have cloned the DNA molecules that code for genes contained with genomes of organisms. Genes that are defective in genetic diseases, and which have been cloned, facilitate the acquisition of new cellular biological and biochemical knowledge that may, in turn, be exploited to treat or circumvent genetic disease, to lessen the suffering connected with genetic diseases. The field of molecular medicine is an active area of investigation.

DNA has practical purposes, too. For example, DNA may be exploited to diagnose diseases, using biotechnological toolkits to do so. Additionally, in forensic medicine, DNA may be used to identify victims and perpetrators of crimes. It may also be effectively used to identify parental lineages in child custody disputes.

In the last several decades, DNA has been used to perform gene therapy, for the specific purpose of correcting a defective gene by replacing it with a properly functioning version of the gene, in order to express the correct gene product properly and reverse the aberrant function that may occur during a genetic disease. Such measures may restore the clinical utility of the normal gene. More recently, a gene-editing technology, called CRISPR, has emerged as the promising new and upcoming invention for correcting a damaged or defective gene in all cells of a living organism.

2) Now, to set the stage for the next scientist, what exactly is RNA and why is it important for scientists?

I think it is intriguing that the traditional DNA structure primarily has one major and very famous 3-D double helical shape, while RNA molecules, on the other hand, have the ability to form a variety of distinctive 3-D shapes. As I briefly alluded to above, RNA serves as an intermediate molecule between the genetic code embedded in DNA and the proteins specified by the code. The three RNA forms, mRNA, rRNA and tRNA, have their individual functions—all of which are geared towards the production of protein chains, which then fold into their respective and functional 3-D shapes.

In terms of the three various types of RNA, the mRNA type, for example, serves as a messenger of the genetic code information. As such, mRNA constitutes a conduit between the code stored in DNA and the biological functions conducted by protein. The tRNA molecules participate in transferring specific amino acids from the cytoplasmic milieu to the translational machinery; and a specific tRNA molecule exists for each of the 20 amino acids that make up protein molecules. Much of the rRNA molecules form a large part of the cellular ribosomes, which are protein-making entities, and they (rRNA) participate directly in attaching amino acids to each other.

With respect to these RNA functions, the investigative works performed in the laboratories of Drs. Sidney Altman and Thomas Cech revealed a new function for RNA. They found that RNA could act like proteins in the sense that they could catalyze biochemical reactions. This was, in fact, a shocking and unheard of discovery. Altman reported that his laboratory had relatively less trouble publishing their findings as Cech's work had been already known.

3) Sidney Altman in the 80's shared a Nobel Prize with Thomas Cech for looking at the catalytic properties of RNA. Why is and was this important?

Prior to each of the Nobel Prize winning discoveries in the research laboratories of Drs. Altman and Cech, RNA had been thought of primarily as a player in the so-called Central Dogma for the flow of genetic information between generations and in the replication of DNA within generations.

As Profs. Altman and Cech independently demonstrated, RNA was found to confer catalytic reactions, similar to those previously known to be performed mainly by enzymatically based proteins. That is to say, Altman and Cech discovered that RNA could act like an enzyme and catalyze a biochemical reaction.

In fact, in order to accommodate the newly discovered function for RNA, the term 'ribozyme' was coined to denote the new fact that RNA could harbor catalytically-based characteristics, similar to those seen in conventional biochemical reactions mediated by proteins. Sidney Altman and colleague Norman Pace studied one of these ribozymes closely in the bacterium called *Escherichia coli*. Referred to as ribonuclease P, or RNase P, the new ribozyme was determined to be a combination of RNA with a protein component, the latter of which was found to stabilize the RNase P ribozyme.

Altman and Pace took an *E. coli* RNase P substrate, a tRNA molecule, and observed that the RNase P had cleaved a portion of the tRNA substrate. The result was that the tRNA product was now a bit shorter than before. Thus, cleaving the tRNA eventually allowed it to become a mature tRNA molecule and now bind to an amino acid, making it ready for participation in the protein-making translational process.

For the first time in recorded history, scientific investigators began to think in terms of RNA molecules serving in the additional new role of mediating biochemical reactions, a process that, until this work by Altman and others had become known, was thought to be performed solely by proteins. The now arcane notion that 'proteins do everything' had been challenged and in a most direct fashion.

4) Apparently, as a youngster two things impressed him—the periodic table of the elements and the dropping of the Atomic Bomb—how did these two things impact his later learnings?

Interestingly, in his brief autobiographical description for the 1989 Nobel, Dr. Altman nostalgically alludes to a gift consisting of a book detailing the periodic table of the elements. Altman conveyed his fascination with the table of elements and the scientific method in general with its elegance and capacity to make effectual new scientific predictions. A few years later, as a teenager, he was to invoke elegantly the scientific method of discovery himself to make one of the most important discoveries in molecular biology and biochemistry.

Dr. Altman spoke about the time he learned about the atomic bomb, designed and developed in Los Alamos, NM, when he was a young child. He had become captivated by the realization that human being scientists had had a direct role in its invention. It brought to the young Altman the feeling that scientific progress could play very important roles in society.

5) For a while, he worked at the Medical Research Council Laboratory of Molecular Biology in Cambridge, England. Why was this important in his professional growth and development?

In an interview, published in 2002, Dr. Altman conveyed his experiences as a fledging postdoctoral fellow at the MRC Laboratory for Molecular Biology. Altman attributes his success in discovering his Nobel-Prize worthy RNase P system to his initial work at the MRC.

Prior to his postdoctoral experience at the MRC, Altman had graduated in 1960 with his undergraduate degree in physics from Massachusetts Institute of Technology (M.I.T.), located in Cambridge, MA, where, as you had briefly alluded, he had become an active participant in M.I.T.'s college hockey team.

Soon after his arrival to the MRC in 1969 to begin a new postdoctoral fellowship, Altman had been initially interested in determining the crystal structure of tRNA. Unfortunately, he had heard through the grapevine by prominent molecular biologists Sydney Brenner and Francis Crick that someone else had already crystalized the tRNA structure.

Thus, Altman decided instead to focus on studying RNA biochemistry in mutational variants of tRNA molecules. The new work led ultimately to his Nobel work.

In order to pursue his interest and conduct these new studies, Altman used a special class of chemical mutagenic compounds, called acridines, to induce the production of mutations in tRNA. Next, he labeled his variant tRNA molecules with a radioisotope to detect his new variant molecules. Then Altman exposed his mutated radioactive tRNA molecules to the cytoplasmic contents of broken bacteria, called a cell lysate; to his astonishment, he found that the front ends of his tRNAs had been clipped off.

He reasoned that there must be some sort of bacterial enzyme activity that chewed off the five-prime (5') ends of his variant tRNA molecules. The mature forms of the tRNA molecules seemed to lack this 5'-end piece.

Now at Yale as an assistant professor, attempts by Altman and his graduate student, Benjamin Stark, to purify the enzyme grew infuriating as the protein failed to make its appearance unless grown in unusually large quantities, and still, an 'RNA contaminant' continually stayed with the protein preparation. It was at this point that Stark had suggested that the enzyme was made of RNA. Being obviously very skeptical at first about the notion of an 'RNA molecule being an enzyme,' they conducted a control experiment in which they destroyed the RNA contaminant with another enzyme, called micrococcal nuclease, known to digest RNA molecules quite effectively.

To their great surprise, the micrococcal nuclease treatment resulted in a loss of the tRNA cleavage of its 5'-end—the RNase P activity disappeared! Next, they repeated their experiment, except this time, they removed a requisite cation, called a calcium ion, which was needed for the RNA-destroying micrococcal nuclease enzyme to function properly.

The calcium removal restored the detection of the putative RNase P enzyme function—they had shown that RNA was an enzyme, and its substrate was tRNA!

In the end, it was shown that RNase P was actually a combination of partly protein and mainly RNA—a two-component ribozyme system. Nevertheless, their experiments definitively showed that an enzyme was RNA.

6) Now, where do ribozymes fit into this big picture? Can you first define them for us and then tell us how they fit in?

Ribozymes are RNA molecules that have the ability to mediate cellular chemistry by catalyzing biochemical reactions. Before the onset of the ribozyme concept, the biochemical catalytic process was primarily the domain of the proteins. The concept that RNA can mediate biochemical catalysis is important for several reasons. Chief among them is the implication that during the first signs of life in primordial seas, RNA-based life may have been the first, rather than DNA. This 'RNA-first' world has important ramifications in investigations of the origin of life on Earth.

Another important implication is that during translation, i.e., protein synthesis, in all living organisms involves a ribozyme type process. In ribosomes, which are known to function as protein synthesizing biological machinery, RNA plays an integral role in translation. In fact, the actual component in rRNA that acutely participates in attaching amino acids to each other, a ribozyme of sorts called peptidyl transferase, has an active site composed of RNA.

One last example considered here is the idea that ribozymes may be used therapeutically to modulate the infective activities of RNA viruses, which may be pathogenic to humans and other animals. This area of scientific investigation is an active one to this day.

7) At the University of Colorado Medical Center, Altman apparently studied the effects of acridines on the replication of bacteriophage T4 DNA. Now what exactly are acridines and why are they relevant to the replication of bacteriophage T4 DNA?

Altman obtained his Ph.D. degree in the field of biophysics at the University of Colorado, housed in their medical center, located in Boulder, CO, in 1967.

It is here in Colorado where Altman had studied effects of the acridines upon special viruses with an affinity for bacteria. Hence, these viruses are known as bacteriophages. The T4 phages that Altman was interested in were already famously known as those that were used in the famous Hershey-Chase experiment to convey the idea that DNA was the heredity material, rather than protein.

The acridine molecules are chemicals with interesting structural and chemical properties. First, these molecules are flat (planar), which permit them to insert themselves in between the nucleotide bases, a process known as intercalation, within a double-stranded DNA molecule.

The acridine intercalation distorts the double helical nature of DNA just slightly enough such that when the DNA replication machinery attempts to synthesize new DNA from acridine-associated DNA templates, the replication machinery will invariably make new mistakes when incorporating new nucleotides. Thus, the acridines play a key role in producing DNA mutations.

Furthermore, as a newly minted Ph.D., Altman had taken a postdoctoral position at Harvard University, in Cambridge, MA where he had worked under a prominent molecular biologist, Mathew Meselson, studying DNA endonuclease enzymes involved in performing DNA synthesis and recombination in the T4 bacteriophage viruses.

8) What have I neglected to ask?

I once had an unprecedented opportunity to meet Dr. Altman. As part of an organizing committee in the early 2000s for a small symposium to be held in Taos, New Mexico for biomedical scientists, I was given the charge of picking up Dr. Altman from the Sunport in Albuquerque and driving him to the conference in Taos, NM, some 2.5 hours away through somewhat rugged, but very pretty, mountainous areas.

Since I had not been accustomed to spending extended periods of time alone with such prominent Nobel Laureates, I was a bit apprehensive about the 2.5 hour drive alone with Dr. Altman. As a young assistant professor still thinking about tenure in the faraway future, I desperately wanted to avoid saying anything that might give him pause about my scientific acumen or me. Any trepidations I had imagined, however, were immediately dispelled from the very moment we drove out of the Albuquerque International Sunport and began our two and a half hour conversation.

We talked freely about science, books, music, and history. It turned out we had been reading similar books; one in particular still stands out in my memory—we had read Frederick Holmes' thick two-volume biography (more like a tome) about the early life and science of Prof. Hans Krebs—Altman was good friends with the author, who was also at Yale.

During the drive to Taos, I took advantage of the unique opportunity and asked Altman many questions. He provided practical research advice that I still follow to this day—he encouraged me to continue my then preliminary studies of bacterial resistance and to be open to new research avenues, should they come up. It was sound advice. It had been clear to me that I had been in the presence of a true genius.

Furthermore, Dr. Altman turned out to be an extremely pleasant and friendly person with whom to hold an extended conversation, and I feel tremendously fortunate to have had the opportunity to pick his brain on that lovely drive through the northern New Mexico mountains.

Chapter 37

THOMAS STEITZ

1) Professor Varela, a recent Nobel Prize winner, shared the prize with two colleagues for their studies of the structure and specifically the function of the ribosome. First, can you tell us exactly what a ribosome is, and why we study it?

I believe that the ribosome has to be the greatest living super biological machine ever invented by living beings. The ribosome is a complex aggregation consisting of intertwining protein and RNA components, whose primary function is to make proteins within the cells of all living organisms.

On a molecular scale, the ribosome is massive. The nature of the prokaryotic ribosome consists of a so-called 70S complex of protein and RNA, the complete structure of which consists of two smaller components, called the 50S and the 30S sub-units.

Incidentally, within the terms 70S, 50S and 30S, the 'S' refers to the Svedberg value or the sedimentation rate ('S' for Svedberg or sedimentation) that results from centrifugation of the various components during their experimental preparation. The S values denote the behaviors of the purified components upon their isolation from living cells. The S values reflect the particular sub-unit's size and shape. They do not add up when individual sub-units are combined with each other because of changes to their in sizes and configurations when combined.

The nature of the eukaryotic ribosome differs from that of its prokaryotic counterpart in the following ways. Ribosomes from eukaryotes are typically composed of an overall 80S protein-RNA complex consisting of smaller 60S and 40S sub-units. The prokaryotic and eukaryotic ribosomes differ sufficiently from each other on a molecular structural basis that, even though they function to biosynthesize protein, their detailed molecular mechanisms of doing so during translation, however, are somewhat distinctive from each other.

The synthesis of proteins, called translation, is one of the primary biological processes that occurs within the insides of all known living cells, from bacteria to humans and every living organism in between. Scientific investigators study its function and structure for many important reasons. Foremost on the list of reasons is that translational machinery is a principal system for making proteins, which in turn carry out many of the processes necessary for organisms to maintain life. In a word, ribosomes are necessary for life.

Another important reason for focusing so much attention on the molecular and functional nature of ribosomes is that they represent a suitable target of anti-bacterial medicines for the treatment of infectious disease. Many of the antibiotics produced by microorganisms target the bacterial ribosome while leaving the human ribosomes largely alone, making prokaryotic ribosomes the prime target for the proverbial 'magic bullet.' At the beginning of the 20th century, Paul Ehrlich devised the magic bullet concept to describe an antimicrobial chemical agent that could selectively target microbial pathogens in the human body without harming the human host itself. Therefore, ribosomes represent one of the principal cellular targets for magic bullets.

Ribosomes are good macromolecules to use, for example, during protein engineering, in order to make products that we may find useful, like medicines for the treatment of a variety of diseases, plus vitamins, foods, diagnostic kits, industrially important molecules, etc. Ribosomes may be used also to produce agricultural products, which help crops to survive insect pathogens, invasive species, or harsh climates. In short, ribosomes represent one of the greatest naturally occurring tools to be used for development of modern inventions.

2) Thomas Steitz, Venkatraman Ramakrishnan and Ada Yonath were awarded the Nobel Prize- but was it for some discovery, or simply their work and investigations?

Simply put, Drs. Steitz, Ramakrishnan, and Yonath earned the Nobel Prize in 2009 in the field of chemistry for their having solved the molecular structure of the ribosome. In essence, taken together, their respective investigative works definitively details the massive molecular nature of the ribosome for prokaryotes.

Dr. Thomas Arthur Steitz had published the molecular structure of the bacterial 50 ribosomal sub-unit. Dr. Venkatraman (nicknamed 'Venki') Ramakrishnan is noted for having solved the 30S sub-unit of the bacterial ribosome.

Dr. Ada E. Lifshitz Yonath solved the entire structure of the ribosome, both sub-units, for bacteria that live in extreme environments, called extremophiles. Interestingly, Dr. Yonath discovered an amazing fact about the bacterial ribosome; she and her co-workers found that it has a tunnel. Presumably, the tunnel is a conduit for the exit of the newly made protein chain of amino acids from the ribosome!

3) Apparently, Steitz and Peter Moore used something called X-ray Crystallography in their work. What exactly is this, and why is it important in scientific investigations?

The X-ray crystallographic method has been used for over a hundred years to determine the detailed molecular structures of important molecules—that is, how the atoms of molecules are spatially arranged in relation to each other. Determining the molecular nature of structures is important because they may reveal their mechanistic secrets in how they function.

A simplified description of the overall process of the X-ray crystallography technique is as follows.

First, investigators convert the molecules into crystals, which may represent repeating units of the structures they wish to determine. During the years since its employment, investigators have realized that certain molecules are easily crystallizable while other are recalcitrant to forming crystals. The crystal forms of molecules are necessary for the method to work.

Second, the crystals are placed in a box within specialized X-ray machines that send X-ray beams to the crystals from all around them. The X-rays then hit the crystals and bounce off of them at certain angles, based on the shapes of the molecular crystals. The bounced (i.e., diffracted) X-rays then hit photographic film that is lined within the box, producing exposure patterns of diffracted X-rays on the films, and the exposed films are then developed.

Third, the exposure patterns of the diffracted X-rays produced on the developed X-ray films are then analyzed. Early on in the history of the technique, the diffraction patterns were evaluated mathematically by hand. More recently, computers do these sorts of analyses. Either way, the molecular structural natures of various molecules are deduced from the X-ray diffraction data.

Lastly, based on the structural information gleaned from the molecules, investigators can readily explore their functions, their mechanisms of action, their gene expression regulation, and their possible modes for modulation. They can make mutational variants that lose function or change their functions such that a new function is created. Often, if a molecule works by performing several steps in its mechanism, it may be necessary to create crystals and determine their structures for each of the steps in the overall process.

4) My gosh, Steitz and Peter Moore, apparently, while at Yale, seemed to have solved some structure of the ribosome - and this event was termed a "groundbreaking achievement like climbing Mount Everest." Was someone exaggerating, or how big an event was this?

The discovery made by Steitz and Moore is about as groundbreaking as it gets. For the very first time in history, investigators got a glimpse of the inner workings of the prokaryotic ribosome. Moore and Steitz elucidated the molecular structure of the 50S sub-unit—the largest ribosomal sub-unit of the ribosome—from *Escherichia coli*, an important organism that's used as a laboratory tool by virtually all molecular biologists.

The work of Moore and Steitz culminated from a massive experimental undertaking. The measurements of the data alone are staggering. The *E. coli* organism is a single cell, and each cell harbors about 20,000 ribosomes within it. An individual ribosome, from a molecular perspective, is colossal, each with a molecular weight of 2.5 million Daltons! It is a super macromolecule.

The *E. coli* 50S sub-unit alone consists of 34 individual known proteins, each with their own distinctive structural configurations and molecular weights, plus two individual RNA molecules, designated as the 5S and 23S rRNA sub-units. Each of the 34 proteins of the 50S sub-unit are designated as L1, L2, L3, and so on.

On the other hand, the 30S ribosomal sub-unit consists of 21 different known proteins, each one designated as S1, S2, S3, etc., and one RNA component, called the 16S rRNA.

Interestingly, prior to the revelation of the ribosome structure, it had been thought that the protein component served as the protein making part, i.e., the actual enzyme that connected amino acids to each other to make protein chains, while the RNA part was thought to be merely structural in its purpose. To the surprise of just about everyone involved in the structural studies and to those interested in proteins and protein synthesis, the opposite was true! That is, the RNA part of the ribosome is the enzyme and the protein part of the ribosome is the structural aspect. In short, the ribosome is a ribozyme.

5) Apparently in Tom Steitz's early years he worked with THE COLONEL- who was actually someone who was from Kentucky named Bill Lipscomb and this Professor Lipscomb had a group of students who were actively working on the crystal structure of bovine carboxypeptidase. Now, I figure that bovine refers to cows, but what is this other word and how does it fit into his research?

William (Bill) Nunn Lipscomb, Jr., had taken his Ph.D. in chemistry in 1946, under the graduate supervision of Prof. Linus Pauling at Caltech. Dr. Lipscomb was later a mentor to Steitz and a variety of other noted scientific investigators throughout the years. Lipscomb was a Harvard professor who is best known for his Nobel Prize winning studies (in 1976) of the boranes, which are molecules of the element boron connected to a variety of hydrogen atoms, called hydrides. Lipscomb's work with boranes helped to determine the nature of the chemical bond.

Apparently, Lipscomb's very first graduate student, Murray King, bestowed the nickname 'Colonel' upon Prof. Lipscomb, who was from KY. Since then, generations of doctoral students and postdoctoral fellows alike have referred to Prof. Lipscomb as 'The Colonel.'

In The Colonel's laboratory at Harvard, Steitz had been a graduate student in the biochemistry and molecular biology doctoral program. For a good portion of his Ph.D.

thesis and in a series of publications, Steitz uncovered the molecular structures of two enzymes called aspartate transcarbamoylase and carboxypeptidase A.

The aspartate transcarbamoylase enzyme catalyzes the biochemical conversion of aspartate plus carbamoyl phosphate to the product carbamoyl aspartate. This biochemical reaction is known as an initial step in the production of pyrimidines, which are precursors of nucleotides for producing DNA down the line.

The carboxypeptidase A enzyme plays a role in the metabolic degradation of certain amino acids by breaking the peptide bonds between amino acids at the end of a polypeptide chain. Carboxypeptidase A is known to play a role in a variety of physiological processes, such as metabolic digestion and blood clotting.

6) X-ray crystallography - what exactly is it, and why is it important? Who first discovered it?

The methodology of X-ray crystallography, or X-ray diffraction, is meant to determine the structural configurations of molecules—that is to say, their shapes. Knowledge of the structural shapes of molecules can often provide important insights about the ways in which they function.

Wilhelm Konrad Röntgen discovered the X-rays themselves in the late 1890s. He earned the Nobel Prize for this discovery alone in the year 1901, in the field of physics. The establishment of the research field of crystallography was instituted by William Lawrence Bragg and William Henry Bragg (William Lawrence's father), and in 1915, they shared the physics Nobel. The developer of the technique in which the X-rays were used to determine molecular structures was Max von Laue, who garnered the Nobel for it in 1914.

The X-ray crystallography was first used to determine structures of small molecules, like various salts and minerals. Proteins and DNA later gave way to X-ray diffraction studies, providing valuable information about their chemistry, their interactions with each other, and in the nature of their functional mechanisms. The technique is widely used to this day.

7) What are the implications of their work today? How did Tom Steitz and Peter Moore and colleagues impact science today?

The solving of the *E. coli* 50S ribosomal structure by Moore and Steitz paved the way for the elucidation of the intact 70S ribosome. The various contact sites between the two sub-units, the 30S with the 50S components, were determined. Furthermore, the binding sites of the messenger RNA (mRNA) within the overall ribosome complex became known.

Investigators were able to distinguish between the P-site (peptidyl site), the A-site (aminoacyl site), and the E-site (exit site) on the ribosome. The P-site on the ribosome binds the first tRNA molecule that's bound to the initiating amino acid, usually a methionine. The A-site on the ribosome represents the place where the rest of the tRNAs bind; these tRNA molecules bring along their attached specific amino acids to this site on the ribosome.

Various locations on the ribosome are target sites for certain antibiotics that inhibit protein synthesis and prevent bacteria from detrimentally taking hold upon a patient, thus thwarting morbidity and mortality. For instance, the 50S ribosome sub-unit has translational inhibiting sites for the antibiotics chloramphenicol, clindamycin, and linezolid, plus members of the macrolides and the streptogramins. These research efforts are geared towards finding new ways to eventually eradicate bacterial antibiotic-resistance, especially in 'superbug' bacteria with multiple-antibiotic resistances.

8) What have I neglected to ask about Tom Steitz and his work with his collaborators?

D. Tom Steitz is married to Prof. Joan Argetsinger, a prominent scientific investigator in her own right. After graduating from university with her degree in chemistry from Antioch College, in OH, Joan A. Steitz had been accepted into the M.D. degree program at Harvard Medical School, in Boston, MA, but turned it down to enter graduate school at Harvard in their prestigious biochemistry and molecular biology program.

Incidentally, Steitz was the first female graduate student in Jim Watson's laboratory at Harvard. With Watson as her graduate advisor, she took her Ph.D. from Harvard in 1967, having studied bacteriophages.

Dr. Steitz holds the prestigious Sterling endowed professorship of molecular biology and biophysics at Yale and is a principal investigator at the Howard Hughes Medical Institute.

Prof. Steitz is primarily known for her scientific discoveries pertaining to the molecular interactions between the mRNA message and components of the ribosome. She found that the contacts between the RNA transcript and ribosome involve complementary base-paring strategies.

Dr. J.A. Steitz also discovered that small nuclear ribonucleotide proteins, called snRNPs (pronounced 'snerps') function in the splicing mode between various RNA molecules to produce mature mRNAs. Her role in discovering these so-called spliceosomes have demonstrated important roles for the snRNPs in the biomedical sciences pertaining to the diagnosis of e.g., lupus, a series of autoimmune diseases.

Chapter 38

KARY BANKS MULLIS

1) Dr. Varela, a scientist named Kary Banks Mullis won a Nobel Prize back in 1993 for supposedly "inventing" what we now call PCR - First of all, what exactly IS PCR (polymerase chain reaction) and while I know basically what a chain reaction is - why is this important?

The polymerase chain reaction (PCR) invention was developed by an extremely fascinating character and noted biochemist Dr. Kary Mullis. The PCR invention is a molecular biological technique that's used to enhance the detection of vanishingly minute traces of DNA; that is, PCR greatly amplifies the amount of DNA, however scarce, that may be present within a given sample, in order to further study the amplified DNA using any one of a variety of molecular technologies. The PCR invention is so powerful that it can detect even one molecule of DNA!

Thus, the PCR procedure provides enough DNA in a sample and allows closer investigation. Often the amount of DNA present in a particular environment is simply too low to conduct any useful analysis by conventional molecular approaches. Therefore, the PCR technique is able to find these minute quantities of DNA and enhance the concentrations by orders of magnitude which permit close and careful examination.

The amplification of DNA is only one benefit of the PCR invention. The procedure may be used to clone novel genes for molecular analysis.

It has been used to determine the sequences of nucleotides along DNA molecules, an early name being given as 'cycle sequencing.' Variations of the PCR technique have been used to sequence entire genomes of organisms.

Another application of PCR has been used to construct various mutations within the sequences of bases encoded in DNA in order to study the effects of the mutations in living cells or tissues or even in entire living organisms.

Furthermore, PCR has been used to identify unknown species or individuals or to determine the parental linage in a custody battle. The invention has been used frequently as forensic evidence in many crime cases.

2) When exactly when this individual born and where did he go to school?

Kary Mullis was born in the small town of Lenoir, North Carolina, USA on the 28th day of the month of December in the year 1944. The Mullis family lived in this area, located adjacent to the so-called Blue Ridge Mountains, until Mullis turned 5 years of age, at which time the family moved to a larger city. Mullis' parents were Cecil Banks Mullis and Bernice Alberta Barker Mullis. Mullis went to Dreher high school, which is situated in the city of Columbia, South Carolina. While in high school, one of his main early interests included the subject of chemistry.

After high school graduation in 1962, Mullis entered the Georgia Institute of Technology (Georgia Tech), located in Atlanta, GA, and studied primarily chemistry as his major area of study. While an undergraduate at Georgia Tech, Mullis worked in a chemistry laboratory devoted to the study of light metal hydrides, such as lithium aluminum hydride. Mullis took his Bachelors of Science undergraduate degree in chemistry from Georgia Tech in 1966.

Another area of study for Mullis was his interest of the biological side of chemistry, a field called biochemistry. Toward this, Mullis entered graduate school in 1966 at the prestigious University of California at Berkeley, CA. Mullis entered the research laboratory of Dr. John "Joe" B. Neilands, professor of biochemistry, and graduate thesis advisor to Mullis. In Neilands' laboratory Mullis initially studied iron metabolism and transport in bacteria with specialized proteins, called siderophores, that would bind iron more tightly than, let's say, human iron-binding proteins, would.

However, Mullis and other graduate students in the research group were given carte blanche by Neilands to study whatever interested them. At first, Mullis took many courses outside of the biochemistry department, except molecular biology—which then became a bit of a problem during his thesis defense. He had enrolled in courses pertaining to music, anthropology, math, physics, and sociology. The story is told that Mullis' journey into these other non-biochemistry and non-molecular biology fields took him into the study of astrophysics, where his first paper was published, in the prestigious journal Nature. This accomplishment was later to prove most helpful in ultimately helping Mullis to earn his Ph.D.

Interestingly, while still in graduate school and enjoying his intellectual pursuits in other disciplines, Mullis had acquired an intense curiosity with the biochemistry department's brand new machine, a specialized spectrophotometer called an NMR, for nuclear magnetic resonance, the equipment of which he used to study the structure of an organic molecule with an interesting name, "schizokinen," related to iron transport systems in a microbe called *Bacillus megaterium*, a rather large Gram-positive bacterium. With a suitable thesis project complete but the problem of having neglected to enroll in molecular biology re-emerging, it became an obstacle to the final thesis defense approval. This is where the Nature paper had come into play—the idea being that if he knew enough of the scientific method that he actually got a paper accepted in Nature, it ought to be considered that his thesis was conducted in much the same sort of way; thus, the wayward publication greatly helped facilitate the final approval of the thesis by his graduate committee members. Mullis took his Ph.D. degree in the field of biochemistry from UC Berkeley in 1972.

3) What does PCR have to do with DNA?

In short, the PCR procedure makes DNA. The amount of DNA produced, however, is enhanced by orders of magnitude, perhaps a million-fold more DNA can be synthesized, or even a billion-fold, if so desired. Furthermore, the DNA production is carried out relatively quickly—within just a few hours, in most cases. The PCR invention is also extremely sensitive, having the capacity to detect extremely tiny quantities of DNA.

Remarkably, in his autobiography tilted "*Dancing Naked in the Mind Field*" Dr. Mullis tells his readers that the idea of the PCR invention came to him almost in a flash while driving in Mendocino County along the California Highway 128 from Berkeley on his way to Anderson Valley, nearby mile maker 46.58. He knew almost immediately that history was about to change as a result of what he had just figured out.

This history-changing invention is described briefly here. The PCR technique itself entails several players in the process.

The first and foremost player is the DNA that is to be detected. This DNA is often called the template or target DNA. The next important player is an enzyme called DNA polymerase, which is a protein that sees the DNA template and uses it to make new progeny DNA. This new progeny DNA is often called the product—it is the end-result of the PCR process and is the entity that's produced in an extremely enhanced manner. Important to this process are the four nucleotide bases, denoted as A, G, C and T, which are free floating in a buffer solution prior to the start of the reaction.

Another critical player is a set of two so-called oligonucleotide primers, which are short stretches of DNA, varying between 10 and 40 nucleotide bases long and which are identical to the two flanking ends of the DNA template and which anneal to these ends of the single-stranded DNA template. The two annealed DNA primers on the template permit the DNA polymerase to start DNA synthesis along the template using the bases in the template as a guide with which to incorporate into the newly made DNA strands along the template.

At its very core, the PCR procedure has three fundamental steps. The first is called template denaturation, but has also been referred to as melting or DNA strand separation.

In this step, the separated DNA strands expose the bases in the template. The second step is called primer annealing, but has also been referred to priming. In this step the two primers bind to the flanking ends of the melted separated DNA strands—one primer binds to the end of one strand of the template, and the other primer binds to the end of the other strand—hopefully not the same strand.

The third step is called elongation, or DNA synthesis. In this stage, the DNA polymerase synthesizes a new strand of DNA along each of the separated primer-annealed template strands, making a double stranded DNA product at the end of the elongation step. Each of the three steps may last only about 15 to 30 seconds or, at most, only a few minutes, depending on the actual length of the overall DNA template molecule.

Then, the three-step process of denaturation, annealing, and elongation, is repeated all over again, making another round of intact double-stranded DNA molecules, but this time, in this second cycle, the end product is doubled over that made in the previously conducted first cycle.

If the three steps (i.e., denaturation, priming and elongation) in a cycle are repeated for about 25 rounds, the original template will have been duplicated about million times over, and if 30 rounds or cycles are performed, then the template will be amplified approximately a billion times the original amount that the investigator had started out with at the beginning of the PCR procedure. In either of these cases, the entire amplification procedure of 25 to 30 cycles may actually take only 2 or 3 hours, and the amount of DNA greatly enhanced, allowing more detailed study via a variety of different ways and for various important purposes.

When Mullis first conducted the PCR process in his laboratory, he merely transferred the test tubes from water bath to water bath, each with their various temperatures set-up beforehand. Nowadays, these PCR cycles with their various biochemical reactions are performed in automated machines that rapidly switch temperatures and do so at their necessary time-periods, making the process more efficient and less cumbersome.

4) Apparently this has something to do with paleobiology? Can you clarify?

The study of paleobiology has to do with the biology of fossil artifacts formed by life forms, such as plants and animals. Before the advent of molecular biological approaches to the field of paleobiology, experts in paleontology relied mainly on the shapes of fossils, such as bone structures, etc., to identify the origins, species, and the evolution of ancient organisms and, for example, the identities of remains of the dead. It was a slow and cumbersome process, sometimes fraught with inaccuracies, due to lack of complete remains or artificial manipulation of fossils and other artifacts.

With the introduction of PCR, however, and the availability of ancient DNA samples contained within ancient fossils, the study of biologically-based artifacts became a simpler endeavor and could provide valuable information regarding identification, function, and evolutionary relatedness between long dead organisms.

Such implementation of the PCR techniques to the study of, for example, ancient dinosaur fossils, sparked the publication of science fiction novels about bringing back the dinosaurs from the dead to the living, as seen in the famous 1990 book *The Jurassic Park* by Michael Crichton and feature film of the same name, with their accompanying sequels.

In his science fiction book, Crichton employed PCR in his scheme to clone the dinosaurs and to bring them back to the modern world as living breathing beings. At a first glance, the process seemed somewhat reasonable. First, the fictitious investigators extracted perfectly intact ancient DNA from dinosaur bone marrow blood preserved in the blood meals of ancient mosquitos encased in amber. Next, the fictional process involved using the PCR to amplify the dinosaur DNA. The so-called genomic dinosaur DNA was then used for gene mapping with restriction enzymes and for nucleotide sequencing. The sequenced DNA was compared with those sequences stored in super large DNA databases.

Any dinosaur genes presumably missing from their genomes were then replaced with modern amphibian DNA—most closely related to dinosaurs, according to Crichton—although actual studies showed that some modern birds have been observed to be most closely related to some ancient dinosaurs. The largely intact dinosaur DNA was cloned, restriction enzyme mapped, injected into host eggs, and allowed to hatch the baby dinosaurs.

In reality, the retrieval of intact genomic DNA for any given dinosaur had encountered several problematic issues. First, ancient DNA was chemically different compared to modern DNA, having been altered with time. Secondly, the extents of the entire genomic dinosaur DNA retrieved was grossly inadequate for cloning at the organismal level. Sure. Perhaps in theory dinosaur genes may possibly be cloned using PCR and other modern cloning methods, but cloning of individual genes are a far cry from obtaining entire genomes for dinosaurs, and certainly not for cloning entire dinosaur organisms. As of this writing no definitive evidence has been presented to indicate that dinosaur DNA has been found. Perhaps molecular techniques may someday improve in other ways, such that recovery of long extinct beings, including dinosaurs, may become possible.

Science fiction aside, PCR is used in modern times for forensics, such as for the identification of an unknown species, or of an individual, such as a crime victim or perhaps a criminal perpetrator. PCR became a household word to the laymen during the mid-1990s as a result of the "crime of the century" case with the O.J. Simpson murder trial. Recently, PCR was used to identify the microbial species of corpses as a function of time, in order to determine the time of death, as the microbial composition on a corpse will be distinctive as the body decomposes. PCR may be used for identifying genes that are thought to be defective in genetic diseases. The invention may also be used to clear up confusion with paternity claims.

Occasionally, PCR is used after exhumation of a person from their grave to solve a mystery. For example, PCR was used to study the exhumed bodies of ancient victims of the plague from the medieval times to definitively determine that the ancient bacterium *Yersinia pestis* was the causative agent of the Great Mortality, also known as the Black Death.

A number of years ago, an idea was proposed in which the remains of President Abraham Lincoln be exhumed from his grave, and PCR would be used to determine whether he had had a genetic disease known as Marfan syndrome, a genetic condition characterized by certain lanky features on the face and limbs. Apparently, permission to carry out the experiment was refused, and the mystery remains.

5) He also has done some things to "mobilize the immune system." How can this be done, why is it important, and how does he go about this?

I imagine you are referring to some of the more recent work, published in 2015, conducted by Dr. Mullis and his research collaborators. In short, his laboratory, as you say "mobilized the immune system" by engineering specialized immune system proteins, called antibodies, to redirect them to seek out pathogenic bacteria in order to target them for destruction.

The immune system mobilization works on the following principles. Mullis and his group noticed that humans do not have a certain antigen, a special sugar that the group calls "α-Gal" which is a triple sugar consisting of two types of galactose sugars plus a modified glucose sugar called N-acetyl-glucosamine, the latter of which is often found in the cell walls of most types of bacteria. The α-Gal triple sugar antigen is lacking in humans and will induce an immune response if found in humans. Exposure to α-Gal in humans will cause them to mount an immune response, making specific anti-α-Gal antibodies.

The Mullis group connected the α-Gal antibody molecule to another molecule called a "DNA aptamer" to form a new conjugate molecule they called an "alphamer." The DNA aptamer is an adapter molecule that has a specific DNA or RNA sequence which is capable of binding to certain bacterial antigens. The alphamer in turn now has the ability to bind the undesirable bacterial antigen and to recruit human antibodies to the site where the pathogenic bacteria are so that the recruited specific antibodies mediate the destruction of these infectious disease-causing bacteria.

In the published work, Mullis and his collaborators were able to mobilize antibodies using their novel alphamer directed against a virulence factor, called M protein, that permits evasion of macrophage-mediated phagocytosis and is made in sufficient quantities by Group A Streptococcus (GAS) microbes, which normally cause a plethora of quite severe diseases. The 2015 work was conducted in Petri dishes (in vitro), and future work is aimed at applying the invention to an in vivo model, such as in laboratory animals and eventually to application in humans.

The work holds tremendous promise in aiding the immune system to fight infections, like GAS and others, in the future.

6) What is he currently working on?

The immune system mobilization described above, represents, as of this writing, some of his most recent work.

Other relatively earlier works include a critical refinement of his PCR technique that changed the course of its usefulness and in a rather captivating way, involving bacteria that live in extreme environments.

The particular refinement process began with a problem, which was that during the PCR steps, the DNA polymerase enzyme itself would melt. Its properly folded amino acid string would unravel, losing its correct shape and ceasing to function in its DNA synthesis function. This prevented the proper DNA amplification process. The solution was found in an extremophile microbe.

That is to say, the usefulness of the invention process for the PCR technique by Dr. Mullis was greatly enhanced due to the ability of certain bacteria to grow and thrive in very high temperatures. The heat-loving (i.e., thermophilic) bacterium called *Thermus aquaticus* (denoted colloquially as Taq) harbors a specialized DNA-making enzyme, called Taq DNA polymerase. This enzyme can function at its best to make DNA at a rather high temperature of 72°C (which is about 160 °F) and can actually resist being destroyed itself by the very high temperatures which are needed to denature the template DNA molecules. Such denaturation temperatures may run as high as 95°C, or over 200 °F. This Taq DNA polymerase heat resistance property allows investigators to easily denature DNA templates and to make new DNA quite efficiently at 72°C.

If using conventional, heat-sensitive DNA polymerase enzymes from moderate temperature-loving organisms to do PCR, then these molecular helpers themselves would be destroyed while trying to separate the double-stranded target DNA templates, making it impossible to successfully amplify the DNA. Using Taq DNA polymerase, however, made this issue go away during the critical heat denaturation step.

7) He was also awarded the Japan Prize- what do we know about this and to whom is it awarded?

The Japan Prize, also known as the Japan International Prize is awarded to individuals who make an outstanding contribution to technology and to science and who have made an extremely significant advancement along the frontiers of new scientific developments.

The very prestigious and lucrative award (over half-a-million US dollars or 50 million yen) was established in the mid-1980s and is funded primarily by the Japan Prize

Foundation and administered by the Japanese Ministry of Technology. The Prize recognizes individual inventors and scientists from all over the world.

In his autobiography, Mullis recounts the award ceremony in 1992 and his enjoyable conversation with the Empress of Japan, whom he colloquially called "sweetie" and inquired whether she had any available Empress girlfriends, all without, apparently, causing any awkward international incidents.

After receiving the Japan Prize, Mullis was approached by a German TV program host, who claimed to accurately predict future Nobel Laureates, and informed Mullis that he was next up for the award in 1992. Mullis began preparing himself for the occasion while trying to be humble about it, while at the same time allowing himself to be the focus of a documentary about his story and his invention. To his dismay, however, Mullis did not actually receive the Nobel in 1992 and actually gave up thinking about it.

All hope was not lost as he was called by Stockholm in 1993, informing him that he had been bestowed the Nobel in Chemistry. He replied "I'll take it!"

Mullis later reported jokingly that receiving the Nobel gave him a ticket into just about any office and a new license as an expert on just about any topic, as long as one does their homework on these new topics beforehand, of course. As a result, Mullis has had several public disagreements regarding, for instance, HIV as a causative agent of AIDS, even though very few of the investigators in each of their fields hold similar views.

8) What have I neglected to ask about this awesome scientist?

Like other awesome investigators before him, and likely after him, too, I surmise, two of the most prestigious journals in the world, Nature and Science both summarily rejected his Nobel Prize worthy manuscripts about his history-making PCR invention.

Likewise, as had occurred with other famous and dedicated scientists who have had their hard work and efforts rejected by journals, Dr. Mullis was quite upset about these rejections for quite some time. Further, like all investigators who are in the same boat, he published elsewhere, this time in *Methods in Enzymology*.

Another controversy involved Mullis' admitted use of the drug called LSD during the 1960s. He apparently became enamored with its use, studying its biochemistry and its physiology, even synthesizing 'legal' variants. He also dabbled with electric shocks upon himself, attempting to control his body's physiology with his mind, even claiming to have accomplished telepathy using his electrotherapy methods. Also, he recounts in his autobiography that while alone in the woods in Mendocino Valley, CA, a raccoon that glowed a green color had spoken to him and that he awoke the next morning not having remembered any events in the interim. He attributed the encounter to possibly a non-earthling alien. No one can argue that Dr. Mullis is not an outrageous character.

REFERENCES

Andersen, J. L., G. X. He, P. Kakarla, C. R. K, S. Kumar, W. S. Lakra, M. M. Mukherjee, I. Ranaweera, U. Shrestha, T. Tran, and M. F. Varela. 2015. Multidrug efflux pumps from Enterobacteriaceae, *Vibrio cholerae* and *Staphylococcus aureus* bacterial food pathogens. *International journal of environmental research and public health* 12:1487-1547.

Atlas, R. M. 2000. *Many Faces, Many Microbes: Personal Reflections in Microbiology*. ASM Press.

Baron, J. 2014. *The Life of Edward Jenner M. D.* Cambridge University Press.

Beck, R. W. 2000. *A Chronology of Microbiology in Historical Context*. ASM Press.

Berg, J. M., L. Stryer, J. L. Tymoczko, and G. J. Gatto. 2015. *Biochemistry*. Macmillan Learning.

Britton, D. R., M. Varela, A. Garcia, and M. Rosenthal. 1986. Dexamethasone suppresses pituitary-adrenal but not behavioral effects of centrally administered CRF. *Life Sci.* 38:211-6.

Brock, T. D. 1975. *Milestones in Microbiology*. American Society for Microbiology.

Brock, T. D. 1988. *Robert Koch, a life in medicine and bacteriology*. Science Tech Publishers.

Brock, T. D., G. L. Geison, and R. J. Dubos. 2014. *Pasteur and Modern Science*. Springer Berlin Heidelberg.

Brownlee, G. G., and E. Southern. 2014. *Fred Sanger - Double Nobel Laureate: A Biography*. Cambridge University Press.

Bruns, M. M., P. Kakarla, J. T. Floyd, M. M. Mukherjee, R. C. Ponce, J. A. Garcia, I. Ranaweera, L. M. Sanford, A. J. Hernandez, T. M. Willmon, G. L. Tolson, and M. F. Varela. 2017. Modulation of the multidrug efflux pump EmrD-3 from *Vibrio cholerae* by *Allium sativum* extract and the bioactive agent allyl sulfide plus synergistic enhancement of antimicrobial susceptibility by *A. sativum* extract. *Archives of microbiology* 199:1103-1112.

Burgos, J. M., B. A. Ellington, and M. F. Varela. 2005. Presence of multidrug-resistant enteric bacteria in dairy farm topsoil. *J. Dairy Science*, 88:1391-1398.

Chernow, R. 2005. *Alexander Hamilton*. Penguin Press.

Cook, G. C. 2007. *Tropical Medicine: An Illustrated History of the Pioneers*. Academic Press, Paris.

Crichton, M. 2012. *Jurassic Park: A Novel*. Random House Publishing Group.

Crick, F. 2008. *What Mad Pursuit*. Basic Books.

Curie, E., and V. Sheean. 2007. *Madame Curie - A Biography by Eve Curie*. Read Books.

Debra, P., and E. Forster. 2000. *Louis Pasteur*. Johns Hopkins University Press.

De Kruif, P. 2002. *Microbe Hunters*. Harcourt Brace.

Diamond, J. 2017. *Guns, Germs, and Steel: The Fates of Human Societies*. W. W. Norton.

Exton, J. H. 2013. *Crucible of Science: The Story of the Cori Laboratory*. Oxford University Press.

Floyd, J. L., Smith, K. P., Kumar, S. H., Floyd J. T., and M. F. Varela. 2010. LmrS is a Multidrug Efflux Pump of the Major Facilitator Superfamily from *Staphylococcus aureus*. *Antimicrobial Agents and Chemotherapy*, 54:5406-5412.

Garrett, L. 1994. *The Coming Plague: Newly Emerging Diseases in a World Out of Balance*. Farrar, Straus and Giroux.

Gaynes, R. P. 2011. *Germ Theory: Medical Pioneers in Infectious Diseases*. ASM Press.

Geison, G. L. 2014. *The Private Science of Louis Pasteur*. Princeton University Press.

Goldsmith, B. 2005. *Obsessive Genius: The Inner World of Marie Curie*. W. W. Norton.

Hager, T. 1995. *Force of Nature: The Life of Linus Pauling*. Simon & Schuster.

Hargittai, M. 2015. *Women Scientists: Reflections, Challenges, and Breaking Boundaries*. Oxford University Press.

Harold, F. M. 1986. *The Vital Force: A Study of Bioenergetics*. W. H. Freeman.

Henig, R. M. 2017. *The Monk in the Garden: The Lost and Found Genius of Gregor Mendel, the Father of Genetics*. Houghton Mifflin Harcourt.

Holmes, F. L. 1991. *Hans Krebs: Architect of intermediary metabolism, 1933-1937*. Oxford University Press.

Jacob, F. 1995. *The Statue Within: An Autobiography*. Cold Spring Harbor Laboratory Press.

Jacobs, C. 2015. *Jonas Salk: A Life*. Oxford University Press.

Jardine, L. 2000. *Ingenious Pursuits: Building the Scientific Revolution*. Anchor Books.

Jardine, L. 2005. *The Curious Life of Robert Hooke: The Man Who Measured London*. Harper Collins.

Johnson, S. 2006. *The Ghost Map: The Story of London's Most Terrifying Epidemic - and How It Changed Science, Cities, and the Modern World*. Penguin Publishing Group.

Joklik, W. K. 1999. *Microbiology: A Centenary Perspective*. ASM Press.

Judson, H. F. 2004. *The Eighth Day of Creation: Makers of the Revolution in Biology*. Cold Spring Harbor Laboratory Press.

Kakarla, P., J. Floyd, M. Mukherjee, A. R. Devireddy, M. A. Inupakutika, I. Ranweera, R. Kc, U. Shrestha, U. R. Cheeti, T. M. Willmon, J. Adams, M. Bruns, S. K. Gunda, and M. F. Varela. 2017. Inhibition of the multidrug efflux pump LmrS from *Staphylococcus aureus* by cumin spice *Cuminum cyminum. Archives of microbiology* 199:465-474.

Kakarla P., Ranjana K., Shrestha U., Ranaweera I., Mukherjee M. M., Willmon T. M., Hernandez A. J., Barr S. R., Varela M. F. 2017. Functional roles of highly conserved amino acid sequence motifs A and C in solute transporters of the major facilitator superfamily. In: *Drug Resistance in Bacteria, Fungi, Malaria, and Cancer.* Springer International Publishing. pp. 111-140.

Krasner, R. I. 2008. *20th Century Microbe Hunters*. Jones & Bartlett Learning.

Kean, S. 2012. *The Violinist's Thumb: And Other Lost Tales of Love, War, and Genius, as Written by Our Genetic Code*. Little, Brown.

Kelly, J. 2006. *The Great Mortality: An Intimate History of the Black Death, the Most Devastating Plague of All Time.* Harper Collins.

Krebs, H., and A. Martin. 1981. *Reminiscences and reflections*. Clarendon Press.

Krebs, H. A., and R. Schmid. 1981. *Otto Warburg*. Clarendon Press.

Kumar, S., He, G., Kakarla, P., Shrestha, U., Ranjana, K. C., Ranaweera, I., Willmon, T. M., Barr, S. R., Hernandez, A. J., and M. F. Varela, 2016. Bacterial Multidrug Efflux Pumps of the Major Facilitator Superfamily as Targets for Modulation. *Infectious disorders drug targets* 16:28-43.

Kumar, S., and M. F. Varela. 2012. Biochemistry of bacterial multidrug efflux pumps. *International journal of molecular sciences* 13:4484-4495.

Kumar, S., Lindquist, I. E., Sundararajan, A., Rajanna, C., Floyd, J. T., Smith, K. P., Andersen, J. L., He, G., Johnson, J. A., Werdann, J. J., Sandoval, A. A., Mojica, N. M., Schilkey, F. D., Mudge, J., and Varela, M. F. 2013. Genome sequence of non-O1 *Vibrio cholerae* PS15. *Genome announcements* 1(1): e00227-12.

Kumar, S., M. M. Mukherjee, and M. F. Varela. 2013. Modulation of Bacterial Multidrug Resistance Efflux Pumps of the Major Facilitator Superfamily. *International journal of bacteriology* 2013:15. doi:10.1155/2013/204141.

Kumar, S., K. C., R., Sanford, L. M., Hernandez, A. J., Kakarla, P., Varela, M. F. 2016. Structural and functional roles of two evolutionarily conserved amino acid sequence motifs within solute transporters of the major facilitator superfamily. *Trends Cell Mol Biol* 11, 41-53.

Lax, E. 2015. *The Mold in Dr. Florey's Coat: The Story of the Penicillin Miracle.* Henry Holt and Company.

Lekshmi M., Ammini P., Kumar S., Varela M. F. 2017. The Food Production Environment and the Development of Antimicrobial Resistance in Human Pathogens of Animal Origin. *Microorganisms.* 5(1):11.

Levy, S. B. 2013. *The Antibiotic Paradox: How Miracle Drugs Are Destroying the Miracle.* Springer US.

Lipmann, F. A. 1971. *Wanderings of a biochemist.* Wiley-Interscience.

Lwoff, A. 2012. *Origins of Molecular Biology: A Tribute to Jacques Monod.* Elsevier Science.

McCullough, D. 2008. *John Adams.* Simon & Schuster.

Miesfeld, R. L., and M. M. McEvoy. 2017. *Biochemistry.* W. W. Norton.

Monod, J., and A. Wainhouse. 1997. *Chance and Necessity: An Essay on the Natural Philosophy of Modern Biology.* Penguin.

Mukherjee, M., P. Kakarla, S. Kumar, E. Gonzalez, J. T. Floyd, M. Inupakutika, A. R. Devireddy, S. R. Tirrell, M. Bruns, G. He, I. E. Lindquist, A. Sundararajan, F. D. Schilkey, J. Mudge, and M. F. Varela. 2014. Comparative genome analysis of non-toxigenic non-O1 versus toxigenic O1 *Vibrio cholerae. Genomics discovery* 2:1-15.

Müller-Hill, B. 1996. *The Lac Operon: A Short History of a Genetic Paradigm.* Walter de Gruyter.

Mullis, K. 2010. *Dancing Naked in the Mind Field.* Knopf Doubleday Publishing Group.

Nelson, D. L., and M. M. Cox. 2012. *Lehninger Principles of Biochemistry*: 6th Edition. Macmillan Learning.

Nicholls, D. G. 2013. Bioenergetics. Elsevier Science.

Olby, R. 2013. *The Path to the Double Helix: The Discovery of DNA.* Dover Publications.

Oshinsky, D. M. 2005. *Polio: An American Story.* Oxford University Press.

Peng, Y., R. L. Hernandez, R. R. Crow, S. E. Jones, S. A. Mathews, A. M. Arnold, E. F. Castillo, J. M. Moseley, and M. F. Varela. 2008. Pasteurized whole milk confers reduced susceptibilities to the antimicrobial agents trimethoprim, gatifloxacin, cefotaxime and tetracycline via the *marRAB* locus in *Escherichia coli. Journal of Dairy Research* 75:491-496.

Prebble, J., and B. Weber. 2003. *Wandering in the Gardens of the Mind: Peter Mitchell and the Making of Glynn.* Oxford University Press.

Ranaweera, I., U. Shrestha, K. C. Ranjana, P. Kakarla, T. M. Willmon, A. J. Hernandez, M. M. Mukherjee, S. R. Barr, and M. F. Varela. 2015. Structural comparison of bacterial multidrug efflux pumps of the major facilitator superfamily. *Trends in cell & molecular biology* 10:131-140.

Ranjana, K. C., Shrestha, U., Kumar, S., Ranaweera, I., Kakarla, P., Mukherjee, M. M., Barr, S. A., Hernandez, A. J., Willmon, M. T., Benham, B. C., and M. F. Varela. 2017. Molecular biology of multidrug resistance efflux pumps of the major facilitator superfamily from bacterial pathogens. In: *Foodborne Pathogens and Antibiotic Resistance.* (O. Singh, ed.), pp. 303-329. John Wiley & Sons.

Ridley, M. 2013. *Genome: The Autobiography of a Species in 23 Chapters.* HarperCollins.

Sayre, A. 2000. *Rosalind Franklin and DNA.* Norton.

Shrestha, U., Ranaweera, I., Kumar, S., K. C., R., Kakarla, P., Lakra, W. S., He, G. X., Andersen, J. L., and Varela, M. F., 2015. Multidrug Resistance Efflux Pumps of *Salmonella enterica*. In: *Salmonella: Prevalence, Risk Factors and Treatment Options* (C.B. Hackett, ed.) pp. 1-30, Nova Science Publishers.

Skloot, R. 2011. *The Immortal Life of Henrietta Lacks*. Broadway Paperbacks.

Smith, K. P., Kumar, S. and M. F. Varela. 2009. Identification, cloning, and functional characterization of EmrD-3, a putative multidrug efflux pump of the major facilitator superfamily from *Vibrio cholerae* O395. *Archives of Microbiology* 191:903-911.

Varela M. F., Andersen J. L., K. C. R., Kumar S., Sanford L. M., Hernandez A. J. 2017. Bacterial resistance mechanisms and inhibitors of multidrug efflux pumps belonging to the major facilitator superfamily of solute transport systems. In: *Frontiers in Anti-Infective Drug Discovery*, pp. 109-131. Bentham Science Publishers.

Varela, M. F.; Wilson, T. H. 1996. Molecular biology of the lactose carrier of *Escherichia coli. Biochimica et Biophysica Acta*. 1276, 21–34.

Watson, J. D. 2011. *The Double Helix: A Personal Account of the Discovery of the Structure of DNA*. Scribner.

Zimmer, C. 2008. *Microcosm: E. coli and the New Science of Life.* Knopf Doubleday Publishing Group.

ABOUT THE AUTHORS

Michael F. Shaughnessy
Eastern New Mexico University, Portales, New Mexico, USA
Michael.Shaughnessy@enmu.edu

Manuel Varela
Eastern New Mexico University, Portales, New Mexico, USA

INDEX

#

A

B

C

D

E

F

G

H

I

J

K

L

M

N

O

P

Q

R

S

T

V

W

X

Y

Z

α

β

λ